Praise for *Darwin and Doctrine*

"*Darwin and Doctrine* provides an invaluable synthesis of science and theology that graciously addresses topics that can often become sources of controversy. I highly recommend it to anyone who wants to see how Catholicism provides a framework to unite faith and reason in the quest to understand the origins of God's creation."

—**Trent Horn**, author of *The Case for Catholicism*

"Many Catholics are unsure how biological evolution fits together with the truths of faith. How do the Genesis accounts of the fall of man fit with scientific discoveries about human origins? Is the role of chance in evolution inconsistent with divine providence? What about the death and violence inherent in the evolutionary process? Daniel Kuebler draws on his expertise as a biologist, his solid grounding in the faith, and his great pedagogical skill to give illuminating and satisfying answers to these and many other questions in this marvelous and much-needed book."

—**Stephen M. Barr**, professor emeritus of theoretical physics and president of the Society of Catholic Scientists

"Despite the questions and assumptions of conflict between evolution and faith, very few books address the issues with the clarity, candor, and insight of this one. Kuebler's love for the Catholic faith and for biological science is palpable. He is a leading voice in the dialogue between the Catholic faith and modern science, and *Darwin and Doctrine* will certainly prove to be his most important and lasting contribution to date."

—**Christopher T. Baglow**, Academic Director, Science & Religion Initiative, McGrath Institute for Church Life, University of Notre Dame

"This is an important book that reveals the compatibility of a Darwinian evolutionary account of the origins of life and the Catholic Church's understanding of God's providential work in creation. It is specifically written for the undergraduate or the educated layperson who is trying to understand both sides of the evolution-creation debate within the rich context of the Catholic intellectual tradition. Of particular significance, this book responds well to many of the objections raised by Catholic creationists who often deploy philosophical arguments to suggest that, in principle, evolutionary change is impossible. In my view, if one accepts that hydrogen and oxygen, two gases, can chemically react with each other to produce water, a liquid, then one should be able to accept that, in principle, two lizards can mate and give rise to a snake."

—**Fr. Nicanor Pier Giorgio Austriaco**, OP, professor of biological sciences and sacred theology, University of Santo Tomas (Manila, Philippines)

"In an age where perceived conflicts with science have led many to feel estranged from the Church, *Darwin and Doctrine* is a master class in Catholicism's 'both/and' vision of the relationship between faith and reason. Drawing from decades of research and teaching experience, Daniel Kuebler presents a comprehensive yet accessible exploration of the most pressing questions at the intersection of creation and evolution. The book speaks to those who feel compelled to reject evolution in defense of their Catholic faith, as well as those who are tempted to abandon their faith, thinking science has debunked Christianity. With remarkable clarity, this must-read book offers a way to understand evolution not as an obstacle to faith but as a testament to the Creator's guiding hand in the natural world."

—**Matthew J. Ramage**, professor of theology and co-director of the Center for Integral Ecology, Benedictine College

Darwin *and* Doctrine

Darwin *and* Doctrine

The Compatibility of Evolution and Catholicism

Daniel Kuebler

Published by Word on Fire, Elk Grove Village, IL 60007

Printed in the United States of America

Cover design, typesetting, and interior art direction by Katherine Spitler, Clark Kenyon, and Rozann Lee

ISBN: 978-1-68578-158-3

Library of Congress Control Number: 2024946117

Contents

Preface

"Then the Lord God formed man from the dust of the ground, and breathed into his nostrils the breath of life" (Gen. 2:7). This line, from the second Genesis creation account, aptly reflects the enigma that is man. As a species, we have our roots firmly planted in the dust of the earth. We are composed of the same stuff as frogs, planets, and stars. The carbon found in our bodies is no different than the carbon formed via the fusion of helium atoms inside stars or the carbon found in a grain of wheat. In fact, it is one and the same. The same carbon that was formed inside a star long ago is now a carbon atom in a grain of wheat. Likewise, that same carbon from a grain of wheat can, through processes of digestion and metabolism, become part of my body. We are fully integrated into the dynamic processes of the physical world.

But that is not the whole story. While we have a home within material creation, we never seem to be fully satisfied within its confines. There is a line from a Bruce Springsteen song that has resonated with me since I was a teenager because it sums up precisely this sentiment. In the song "Badlands" he sings, "Poor man wanna be rich, rich man wanna be king / And a king ain't satisfied 'til he rules everything." But we can never rule everything—life makes that abundantly clear—so we are inevitably restless. There is a transcendent aspect to our being that makes us recognize that there is a longing this material world fails to satisfy. It is why Augustine exclaims to God in his *Confessions*, "Our hearts are restless till they rest in Thee."[1]

This dichotomous union of dust and breath, of worldliness and transcendence, is what Scripture tells us man is. But what

1. Augustine, *Confessions*, trans. F.J. Sheed, ed. Michael P. Foley (Park Ridge, IL: Word on Fire Classics, 2017), 2.

exactly does that mean? Is the dust merely a shell for our true self, that inner spirit? Or is the spirit merely an epiphenomenon tacked on to our true animal selves, selves that are fundamentally the same as other animals? How do we truly integrate matter and form, spirit and matter, soul and body? How do we integrate what the science reveals about the human body with what we know through faith and reason about the human person? These are the questions that lie at the heart of the evolution and creation discussion, and these are the questions that have doggedly followed me ever since I began to reflect seriously on evolution as an undergraduate student.

Over the years, my thinking about the science of evolution has itself evolved significantly, from the cursory simplistic understanding of Darwinian evolution I had as an undergraduate to a more complex and nuanced awareness of the broader extended evolutionary synthesis (EES). Likewise, my thinking regarding how evolution might fit within a Catholic understanding of both creation and man has developed over the years. At the beginning of this journey, I was merely exploring whether I could fit the two together without doing violence to either Catholic teaching or the scientific evidence. But this book is less about avoiding conflict between the two—although that is certainly an important aspect—and more about how evolution and Catholic teaching can be mutually enriching and complementary. Over the years, I have gone from wrestling with how evolution might fit with Catholicism to exploring how an evolutionary understanding can enlighten our understanding of how God relates to his creation.

It took quite some time for me to arrive at this stage; in fact, when I initially began this book during an academic sabbatical in 2014, I wasn't quite there yet. As I reread that first draft at the end of my sabbatical, I realized that I was not quite prepared to write the book I had hoped to write. Not sure what to do and having to return to full-time teaching, I put the draft of the book aside, and there it sat, undisturbed, until I finally had the time to turn my attention to it again six years later.

While I did no writing or editing of the book during that six-year period, it was precisely during those years that the book in its present form took shape. There were three things that happened during that time that helped me develop my thinking on evolution and Catholicism such that this book became possible. All three of these experiences put me in regular contact with colleagues who, through their wisdom and efforts, helped me gain a much deeper and richer understanding of the issues regarding evolution and creation.

The first experience involved being asked to create a course on science and Catholicism for our Catholic Studies master's program at Franciscan University. As I began to develop and then teach the course, I recognized how much I needed to augment my understanding of the Catholic theological and philosophical tradition. On this front, I found that my colleagues in these disciplines at Franciscan University were more than willing to share their expertise. It was conversations, recommendations, and fraternal corrections from these colleagues that particularly deepened my understanding of the Catholic intellectual tradition. As the book took shape, my Franciscan colleagues Paul Symington and Brandon Dahm from the philosophy department provided much-needed feedback on the chapters that were heavy in philosophical content (at least heavy for a biologist), and for that I am most grateful.

The second pivotal event was the foundation of the Society of Catholic Scientists in 2016. In the SCS, I was blessed to find a group of serious thinkers who were looking to integrate science and Catholicism in all aspects of their lives. From interacting with the speakers at our annual conferences to the after-dinner conversations I had with colleagues about such topics as original sin, the nature of creation, life on other planets, and Adam and Eve, my interactions with fellow SCS members have enriched me both personally and intellectually. In particular, Steve Barr, the president of the SCS, has been a constant source of guidance and support during the writing of this book. It was his encouragement years ago that got me started down the path of producing it, and his willingness to share his incalculable breadth of knowledge on science and faith topics has been invaluable to me. Over the years, he has been generous enough with his time to read early chapters of the book, and every editing suggestion or additional insight he provided has vastly improved the final version. Most importantly though, it is his friendship and the conversations, both serious and humorous, we have had over the years for which I am most grateful.

The third pivotal event that happened during this time was an invitation to become involved with the Science and Religion Initiative at Notre Dame's McGrath Institute for Church Life. The director of the initiative, Chris Baglow, asked me to assist in their efforts to help high school teachers better integrate Catholicism and science in their classrooms. Participating in these events has been career-changing. The colleagues I have met through these seminars and the interdisciplinary discussions we have had over beers and cigarettes (even though I don't smoke) have been invaluable for both my intellectual formation and the completion of this book. While I have been blessed to meet and learn from innumerable colleagues and participants during these seminars, it has been the repeated discussions with Chris Baglow, Heather Foucault-Camm, Cory Hayes, and Jordan Haddad that have particularly influenced my thinking and my writing on evolution and the faith. I am very grateful to Cory for the time he took to edit chapters and enlighten me on critical theological and philosophical points that, through his wisdom, eventually made their way into the book.

As I worked through the final version, Joe Miller, my collaborator on the Purposeful Universe project (funded by the John Templeton Foundation), was invaluable. Joe's encouragement, feedback, and stimulating phone conversations provided insightful edits and helped get the book over the finish line.

I am most grateful to Word on Fire Publishing for their interest in the manuscript as the final version took shape. While it is a topic that often sparks controversy and acrimonious discussions, it is nevertheless an important one for Catholics to reflect upon seriously. My editor, Matthew Becklo, has been supportive throughout the entire process and has not shied away from including some of the more speculative sections of the book. He has made excellent suggestions to sharpen and focus the manuscript, and the final version that has emerged owes much to his efforts.

Finally, I would like to thank my wife, Nellie, not only for her constant love and support, but also for her intellectual contributions to this book. Most of the original ideas found in it emerged directly from conversations the two of us have had, and they flow from insights she gleaned or theological connections she made—connections that have changed the way I view the issue. In addition, she has been the first and second (and often third) editor of the manuscript, striking out much of the academic jargon, reordering sections for clarity, and generally making it readable. Any mistakes or issues that remain are solely due to my inherent stubbornness rather than her excellent editing skills.

Clearly this book would not have been possible without the influence of a host of people who have found the issue of creation and evolution as fascinating as I do. Even my children, Joseph, Patrick, Carolyn, Brendan, Julia, and Elena, have contributed. Their curiosity on the subject and the pertinent questions they posed forced me both to rethink my ideas and to learn to express them more clearly. Their curiosity and support have not only helped bring this book finally to fruition, but also made me acutely aware of what I still do not know on the subject.

Toward that end, the reader will notice that there are many questions posed in the book that go unanswered. In addition, there are other questions for which I propose tentative answers, recognizing that in many cases the truth on these matters may be forever beyond our grasp. But this does not make the exercise of exploring these questions fruitless. While acknowledging that some aspects of the creation/evolution discussion transcend our understanding, by using faith to enlighten our reason, we can hopefully come closer to the truth on these matters. Along these lines, it is my hope that this book helps readers come closer to the ultimate truth about creation and evolution, and aids them—as they seek understanding enlightened by the richness and depth of the Catholic faith—in asking the right questions.

1

Introduction

I mean that it is as strange that monkeys should be so like men with no historical connection between them, as the notion that there should be no course of history by which fossil bones got into rocks.[1]

—St. John Henry Newman

"How exactly do you make sense of the story of Adam and Eve given the science of evolution?" That wasn't the question I was expecting my friend to ask when we sat down to eat lunch in the shade of the eucalyptus grove outside our lab building. He was a post-doctoral researcher, someone whom I respected and who had helped me get my bearings as a young graduate student in molecular biology at the University of California, Berkeley. He knew I was both a practicing Catholic and a budding scientist, and he was genuinely curious to get my opinion. It seemed to him that you had to choose only one option: either creation or evolution.

I could certainly see where he got that impression. Unlike most scientific theories, evolution had been co-opted in the larger cultural wars, with many people using evolution as the trump card in the battle of science over religion. Particularly in Berkeley at the time, one couldn't walk more than two blocks without spotting a car proudly displaying the Darwin fish sign on its bumper. The Darwin fish was a response to the Christian fish logo that represented Christ. It was an outline of a fish that not only had the word Darwin inscribed within it but had evolved

1. John Henry Newman, Letter of December 9, 1863, in *Sundries*, 83, cited in A. Dwight Culler, *The Imperial Intellect* (New Haven, CT: Yale University Press, 1955), 267.

legs as well. An even more polarizing version had the Darwin fish eating the Jesus fish.[2]

The Darwin fish was a fitting symbol that succinctly summed up what the atheist philosopher Daniel Dennett deemed to be the logical consequences of evolution. He referred to evolution as a "universal acid," an idea that necessarily alters our way of thinking about everything—largely by eating away at our belief in a loving providential Creator. It was a visual representation of the biologist Richard Dawkins' famous claim that Darwin's theory allowed him to be an intellectually fulfilled atheist. As a lifelong committed Catholic, I was confident that the science of evolution did not necessarily lead to atheism or override the truths of the faith. But despite this confidence, I wasn't sure how to answer my friend's question.

By the time I had arrived at UC Berkeley to pursue my PhD, I had eighteen years of Catholic schooling under my belt. In addition to an undergraduate English literature degree from the Catholic University of America (CUA), I had also completed a master's in cell biology from that same institution. As an undergraduate at CUA, I had been introduced to many of the great minds of the Catholic tradition, including Aquinas, Anselm, Bonaventure, and Augustine. As a graduate student at CUA, I had seen science and religion not only coexist but flourish. The notion that Catholicism was anti-intellectual or that the science of evolution and Catholicism were somehow in conflict was completely contrary to my lived experience at CUA. Here, though, sitting across from my inquisitive friend in the warm noonday sun of Berkeley, the apparent conflict between evolution and my faith was staring directly at me, awaiting an answer.

2. A new Christian version in this symbol war is a fish labeled Truth eating the Darwin fish.

THE EVOLUTION QUESTION

In my twenty-two years of teaching at Franciscan University of Steubenville, I have encountered many students wrestling with the same apparent conflict between the science of evolution and the Catholic understanding of creation. These students have approached me with the same question my friend in Berkeley asked, as well as many others besides. How can an evolutionary process that is influenced by chance events be reconciled with the belief that God deliberately and purposefully created human persons in his image? If evolution is correct, how do we properly understand the Genesis account of the origin of the universe and of man? How can a naturalistic evolutionary process bring about the human person who is at once spiritual and material? How can man be viewed as having been created in a state of original justice and holiness if evolution indicates the world has a four-billion-year history of death and extinction?

These are all key questions that Catholics struggle with regarding the evolution/creation issue. What makes all these questions particularly difficult is that they sit at the intersection of theology, philosophy, and science. Science alone cannot address them, nor can theology alone. These two disciplines, with the aid of philosophy, must be in dialogue with each other to successfully address these issues. Unfortunately, such dialogue is a particularly difficult task in our increasingly fragmented world.

Yet, to move the dialogue forward, an interdisciplinary approach is essential. Clearly, one must properly address the scientific questions that impact the debate. What exactly does the scientific data demonstrate about evolution? Is it a solid theory? What data suggests that it applies to the origin of man? However, this is not enough. One must properly address a host of theological questions as well. What does the Church teach about the creation of man? What does the Catholic understanding of creation entail? What does the Church teach about the proper exegesis of Genesis? While addressing these scientific and theological questions is critical, alone that is not enough. In fact, the most difficult task of all still remains: integrating the truths from both science and theology into a cohesive whole.

Fortunately, it is precisely in this task, the integration of knowledge, where the Catholic intellectual tradition redounds to our benefit. Despite often being lambasted as backward and irrational by the popular culture, the Church boasts a rich intellectual heritage, having spent two millennia systematically addressing fundamental questions that integrate knowledge from all disciplines of study. Throughout its history, in its documents and in its Doctors, the Church has examined fundamental questions related to human existence and meaning, the relationship between science and faith, and the nature of God and creation. For those approaching the thorny scientific, theological, and philosophical issues in the creation/evolution discussion, they could not find another guide with more experience dealing with these big questions.

Even when it does not have answers readily at hand to some of the novel questions posed by evolution, the Church, with its deep intellectual tradition, still provides a productive framework in which to properly investigate them. It is a framework that does justice to both the relevant science and the theology. In this discussion, neither negates nor subsumes the other. As Pope St. John Paul II has indicated, "Both religion and science must preserve their autonomy and their distinctiveness. . . . The unprecedented opportunity we have today is for a common interactive relationship in which each discipline retains its integrity and yet is radically open to the discoveries and insights of the other."[3] Investigating the evolution/creation question within this framework, all under the guidance of the Church, is the purpose of this book.

3. John Paul II, "Letter of His Holiness John Paul II to Reverend George V. Coyne, SJ, Director of the Vatican Observatory," June 1, 1988, vatican.va.

WHAT DOES THE CHURCH SAY ABOUT EVOLUTION?

One of the most common questions that my students ask me regarding evolution is the following: "What is the Church's position on evolution?" This seems to be the million-dollar question that Catholics have regarding the evolution/creation debate. Interestingly, the guidance provided by the Church on this matter is not as straightforward as one might expect. In fact, when it comes to a position on the science of evolution, it turns out that the Church *doesn't* have one. Just as the Church doesn't have an official position on atomic theory or the theory of general relativity, the Church doesn't have an official position on the *science* of evolutionary theory. This should not be surprising given that the Church is not in the business of adjudicating the validity of purely scientific theories. Cell theory, the theory that all living organisms are composed of cells, will rise and fall on the scientific evidence, not on some papal decree. Now, if a scientific theory oversteps its bounds and begins making philosophical claims—as evolutionary theories often do—the Church has every right and indeed the obligation to weigh in on such matters, as it properly has regarding evolution over the years. As John Paul II stated, "The church's magisterium is directly concerned with the question of evolution for it involves the conception of man: revelation teaches us that he was created in the image and likeness of God."[4]

However, it is important to remember that the Church's focus is not on the different *scientific* theories of evolution. Rather, as John Paul II explained, it is concerned about the pseudo-scientific "theories of evolution which, in accordance with the philosophies inspiring them, consider the spirit as emerging from the forces of living matter or as a mere epiphenomenon of this matter."[5] Such philosophical claims are incompatible with what revelation tells

4. John Paul II, "On Evolution" 4, Message to the Pontifical Academy of Sciences, October 22, 1996, in *Origins* 26, no. 25 (December 5, 1996): 415.

5. "On Evolution" 5 (emphasis added).

us about the truth of man. It is these philosophical claims regarding the science of evolution that explicitly concern the Church, not the scientific theory itself. The science must stand or fall on its own weight.

As a result, when it comes to accepting the *science* of evolution, faithful Catholics are free to hold a range of positions on the matter. Thus, there are Catholics who reject evolution outright and see it as incompatible with a historical reading of the Genesis creation accounts and/or Thomistic philosophical principles. At the other extreme, there are Catholics who believe that the evolutionary process, over a four-billion-year timeframe, has produced the type of creature that was fitting to be endowed with a rational soul—namely, humans.[6] Others accept the evolution of nonhuman life forms but hold to the position that humans were created in a direct manner from the dust of the earth and the breath of God as described in the second chapter of Genesis. Still others accept some aspects of evolution but argue that God directly intervened to alter the material evolutionary process at certain points, a view that falls within the spectrum of views held by the modern-day Intelligent Design movement.[7]

So, while the Church has not and will not take a definitive position regarding the various views on the *science* of evolution, this in no way implies that the Church has nothing of importance to say on the matter. Rather, the Church, through official documents and papal writings, has provided a framework to guide Catholics in their approach to the evolution/creation question. For starters, the Church has repeatedly made clear that the science of evolution cannot explain the totality of man. Even if one is convinced by the scientific evidence that man evolved

6. These individuals are often called theistic evolutionists. Given the diversity of opinions that are lumped under this term, it is not a very useful categorization for the purposes of this book.

7. Like the term theistic evolutionist, there are a multitude of positions that are lumped under the broad classification of Intelligent Design. Dr. Michel Behe's work (see *Darwin's Black Box*) is most associated with this position, but the term Intelligent Design encompasses a broad range of positions, making it difficult to define succinctly.

from other primates, the origin of man involves the immediate creation of an immaterial soul, something no scientific theory can explain. The Church also makes clear that the evolutionary process is entirely dependent upon God and does not operate independently from him or his divine will. Regardless of the exact natural processes by which evolution has operated, even if these have involved chance events, it operates in precisely the manner God has intended from all eternity.

Much more will be said regarding these points in later chapters, but the key notion to address in this opening chapter is that while the Church allows a range of positions regarding the science of evolution, there are some philosophical conclusions that are often appended to the science of evolution that are *not* compatible with Catholic theology.[8] Obviously, any atheistic interpretation of evolution is incompatible with the faith. So too, though, is a reading of evolution that assumes God is "surprised" by the outcome of evolution, that it operates independently from his plan for creation. Both of these positions are *philosophical* in nature and go well beyond what the science can demonstrate. They are also clearly in conflict with the teachings of the Church. This is why the Magisterium has spoken out repeatedly against these erroneous "evolutionary" claims for the past hundred years.

A GUIDE MAP TO THE BOOK

While the Church allows faithful Catholics to hold many different positions on the science of evolution, this does not mean that all such positions described above are equally valid. In fact, only one of the scientific positions can be correct. Either natural processes can explain the physical evolution of organisms or they cannot. Either evolution can explain the origin of the man's physical form or it cannot. One of these positions must be closest to the truth, and seeking the truth matters, particularly given that the truth

8. This will be discussed in more detail in chapter 2.

about this topic impacts both our understanding of God's creation and our understanding of God's Word in Sacred Scripture. If the science of evolution is correct about the emergence of new life forms over time, this reveals something about the relationship between God and his creation. If the science of evolution is true, it also reveals something important that impacts the correct exegesis of the first chapters of Genesis. The truths science reveals about the natural world must align with the truths we learn from Scripture because, as Pope Leo XIII wrote in *Providentissimus Deus*, "truth cannot contradict truth"[9] given that all truth comes from one source, the triune God.

Thus, there can only be one perspective that holds to the truths of the faith *and* is consistent with what actually happened during the course of natural history. And if the truth is worth seeking (hint: it is), then we have an obligation to seek it. This is not an easy task though, particularly given the tentative nature of science and the incomplete understanding we have of the mysteries of the faith.[10] Pope Leo XIII's words regarding the caution and care needed to discern the proper interpretation of Scripture apply aptly to the difficulties of discerning the truth at the heart of the evolution/creation discussion:

> Judicious theologians and commentators should be consulted as to what is the true or most probable meaning of the passage in discussion, and the hostile arguments should be carefully weighed. Even if the difficulty is after all not cleared up and the discrepancy seems to remain, *the contest must not be abandoned*; *truth cannot contradict truth*, and we may be sure that some mistake has been made either in the interpretation

9. Leo XIII, *Providentissimus Deus* 23, encyclical letter, November 18, 1893, vatican.va.

10. While science is tentative, there are things we can know with relative certainty via the scientific method. However, new knowledge can always lead to revisions of that knowledge. In terms of our understanding of the mysteries of the faith, we will never completely grasp them precisely because they are mysteries. This does not mean we cannot have some reasonable knowledge of them, but it does mean we can never fully grasp their riches.

> of the sacred words, or in the polemical discussion itself; and if no such mistake can be detected, we must then suspend judgment for the time being.[11]

Pope Leo XIII was not alone in exhorting scholars to engage with these difficult questions. Pope Pius XII, in *Humani Generis*, encouraged that "in conformity with the present state of human sciences and sacred theology, research and discussions, on the part of men experienced in both fields, take place with regard to the doctrine of evolution."[12] Likewise, Pope St. John Paul II and Pope Benedict XVI actively encouraged, supported, and took part in such discussions during their respective pontificates.

In my own life, I have been blessed to play a small role in this interdisciplinary work and to realize some of the fruit that comes from the "research and discussions" envisioned by the Church on this matter. As a faculty member at Franciscan University and through my work in the greater Catholic and Christian academic community, I have been able to participate in regular discussions (both formal and informal) with colleagues in theology, philosophy, history, and science regarding evolution and the faith. I have had the good fortune of learning from Scripture scholars, patristics experts, Thomistic philosophers, personalist philosophers, moral theologians, and Church historians. This is in addition to the many stimulating conversations I have enjoyed with fellow members of the Society of Catholic Scientists, conversations that broadened my scientific knowledge beyond my narrow area of expertise. Without the assistance, guidance, expertise—and most importantly fraternal correction—from these colleagues, this book, which draws upon a wide range of scientific, philosophical, and theological knowledge, would not have been possible.

11. *Providentissimus Deus* 23 (emphasis added).

12. Pius XII, *Humani Generis* 36, encyclical letter, August 12, 1950, vatican.va.

THE GOALS OF THE BOOK

The very name *uni*versity, an institution borne from the heart of the Church, belies the Christian belief in the unity of all knowledge. While all disciplines have their unique methods and subject matter, they are all meant to be rowing in the same direction. Each is meant to discover what is good, true, and beautiful in its own discipline, a revelation that moves man closer to the ultimate source of goodness, truth, and beauty. The picture gleaned from any one discipline, though, is limited and incomplete, like a mosaic produced with only one color of tile. To glimpse reality in all its splendor, one must attempt to properly integrate the truths revealed from all the disciplines.

This interdisciplinary work is immensely difficult, though, and requires significant collaboration as well as intellectual humility. To make matters worse, everyone who enters this dialogue carries with them biases and prejudices. When it comes to the evolution/creation discussion, everyone, this author included, carries cognitive, scientific, metaphysical, educational, and religious biases and preferences that affect how they view and interpret the data. None of us come with a purely objective view or what the philosopher Thomas Nagel called "the view from nowhere." This is not to deny that there is an objective reality; it is simply an acknowledgement of our common human condition, a condition that can hinder our collective search for the truth.

For the benefit of the reader, it is important that I lay out my position regarding creation and evolution here at the beginning of the book because it will be the lens through which I interpret and discuss the data. I believe that the scientific evidence for the evolution of life over the past roughly 3.8 billion years on Earth is robust. I believe natural processes—natural selection plus other natural processes that will be discussed in later chapters—can explain the evolutionary changes that have occurred on our planet. However, I believe, for both philosophical and theological reasons, that mankind is more than a material being. As such, the origin of mankind cannot be fully explained by a material evolutionary process but required the direct intervention of God in the creation of a man's immaterial soul.

In addition, I believe that God created a highly ordered world, a rare universe in which evolution via natural processes is indeed possible. This order, which has been gifted to our universe, is essential for evolution to work at all and can also affect the direction and outcomes of the evolutionary process. While there is order at the heart of the evolutionary process, chance still plays an important role. It is through this dynamic interplay of order and chance that, in the fullness of time, evolution produced the physical entity that was fitting to be informed by a rational human soul, man made in God's image.

While I believe this position is reasonable and can be supported scientifically, theologically, and philosophically, I recognize it is not the only logically possible conclusion. I also realize that my position, like every other position in the evolution/creation debate, raises several sticky issues, from the interpretation of Genesis to the nature of original sin. I certainly do not claim to have the answers to all these questions. In fact, as Pope Leo XII stated, maybe "we must . . . suspend judgment for the time being" on some of them. However, this does not diminish the value of partaking in serious reflection upon these questions.

In my twenty-plus years of teaching at Franciscan University, I have encountered a good number of students who have told me that they reject evolution because they believe it to be incompatible with Catholicism. I have also had others come talk to me about friends and family members who fell away from the faith, in part because they felt the science of evolution had completely undermined Christian belief. This book is largely written with these two populations in mind: those who reject evolution because they believe it inherently conflicts with the faith and those who have jettisoned their Catholic faith because of the belief that it is incompatible with the scientific truths of evolution.

While I have outlined my specific position regarding the evolution/creation question, my primary goal with this book is not to convince readers of the validity of my position. Rather, the goals I have in writing the book are much more modest. My first goal is to provide the reader with a framework that allows them to see how the science of evolution can be reconciled with the truths of the faith. Even if the reader still has scientific concerns regarding evolution after reading the book, if he or she walks away with an understanding that one does not have to choose between evolution and Catholicism, I deem my efforts a success.

If one no longer thinks he or she must demonstrate that natural evolution is false in order to save appearances for Catholicism, he or she is much more able to learn about evolution on fair terms. This leads into the second goal of the book: to provide the reader with a proper understanding of the science of evolution. Unfortunately, students often come into my class with a rather one-sided understanding of evolution. They tend to view it as merely a chance process, one in which random mutations do all the work and organisms are passive bystanders. The reality is quite different, as evolution is built upon a bed of order and shaped by active living agents. The order in the cosmos—the order described by modern physics—is an essential gift to evolution that allows it to operate and helps "direct" it toward certain biological solutions. In addition, biological systems operate in ways that aid and abet the evolutionary process. This orderly view of evolution, although it does not discount the chance elements of the process, leads to a more accurate and awe-inspiring view of that process.

CREATION AND EVOLUTION

When people disagree with certain scientific data, you often hear them discount this data with the claim that the science is always changing and that new data may totally upend the way we think about certain scientific issues. Many proffer this as an excuse to discount our current understanding of evolution. Why try to integrate Catholic teaching with a scientific theory that will be obsolete fifty to one hundred years from now? This type of reasoning, though, is an example of not seeing the forest for the trees. Science is indeed always providing new data for us to integrate into our understanding of the world, and of course evolutionary science is no different. Particularly when it comes to the science of human evolution, the data we have today is constantly being updated and revised. However, while many of the details about evolution will certainly change, the broader theory of evolution is not going away and is about as likely to be discarded by future scientific discoveries as the theory of gravity. In fact, there are many big picture generalities regarding evolution and human evolution that are extremely well supported by multiple lines of evidence. The question is how best to make sense of this evidence in light of a Catholic understanding of both creation and the human person.

Like many things in Catholic theology (and science for that matter), I believe the best way to make sense of the evolution/creation question is to view it as a both/and situation rather than an either/or situation. Just like Christ is both human and divine, just like light displays properties of both a wave and a particle, the truth is that humans have both evolved *and* been directly created by God. The rest of the book attempts to unpack what exactly this means.

2

Evolution and Creation

The Unity of Faith and Reason

The theory of evolution does not invalidate the faith, nor does it corroborate it. But it does challenge the faith to understand itself more profoundly and thus to help man to understand himself and to become increasingly what he is: the being who is supposed to say Thou to God in eternity.[1]

—Joseph Ratzinger

In the 1980s, Cardinal Joseph Ratzinger (who would go on to become Pope Benedict XVI), gave a series of four homilies devoted to unpacking the riches of the first chapters of Genesis. Not long after, these homilies were compiled into a short book entitled *"In the Beginning . . .": A Catholic Understanding of the Story of Creation and the Fall.* While the subject matter of this book is quite expansive, one highlight of the book is the productive framework that Ratzinger laid out for Catholics seeking to integrate evolutionary theory and Catholic theology. Rather than seeing the evolutionary account and the Genesis creation account as two competing theories regarding man's origin, Ratzinger points out in the third homily of the book that evolution and creation are "two complementary—rather than mutually exclusive—realities."[2] He argues that while both accounts provide information regarding man and his origins, they do so on different levels. One offers a scientific account of the relationship between man

1. Joseph Ratzinger, "Schopfungsglaube and Evolutionstheorie," in *Wer ist das eigentlich—Gott?* ed. H.J. Schulz (Munich: Kösel, 1969). Reprinted in *Creation and Evolution: A Conference with Pope Benedict XVI in Castel Gandolfo*, ed. Stephan Horn and Siegfried Wiedenhofer (San Francisco: Ignatius, 2007), 16.

2. Joseph Ratzinger, *"In the Beginning . . .": A Catholic Understanding of the Story of Creation and the Fall* (Grand Rapids, MI: Eerdmans, 1995), 50.

and other primates, while the other offers a theological account of both man's relationship to the Creator and his ultimate destiny. Thus, each account offers a perspective that the other lacks, and it is only when both accounts are properly understood that one can gain an integrated understanding of man and his origin. From Ratzinger's perspective, instead of competing against each other for dominance, the science of evolution and the theology of creation can partner to provide a wider, wiser, and deeper account of reality.

Ratzinger's position that science and Catholic theology are partners in the search for truth is not a novel position. Rather, it is one that has deep roots in the Catholic tradition, a tradition that has produced such scientific luminaries as St. Gregory the Great, Copernicus, Galileo, Mendel, and Lemaître. Yet, despite this impressive track record, the default cultural stereotype often assumes that science and Catholicism (or theology in general) are warring combatants. It assumes that science and faith are two disparate ways of viewing the world, with the former relying on reason and logic and the latter relying on superstition and myth. In fact, recent surveys have found that over half of young people who leave the Church cite the incompatibility with science as one of the factors in their decision.[3] For these young people (and many others), it is science that holds the answers to all the pertinent questions that face humanity, while faith seems to be an archaic and increasingly irrelevant system of beliefs that is slowly losing any influence over our increasingly technology-driven society.

In this cultural milieu, it seems quite ironic that Pope Benedict could proclaim, in a 2009 visit to Africa, that "the task of religion today is to unveil the vast potential of human reason."[4] Benedict's view that religious faith holds the key to unlocking the "potential

3. Saint Mary's Press Catholic Research Group and the Center for Applied Research in the Apostolate, *Going, Going, Gone: The Dynamics of Disaffiliation in Young Catholics* (Winona, MN: St. Mary's, 2017).

4. Benedict XVI, "Pope's Words to Cameroon Muslim Leaders," May 19, 2009, https://www.catholicculture.org/culture/library/view.cfm?recnum=8830.

of human reason" runs counter to the slow atrophy of religious belief in the Western world. For those like the scientific atheist Richard Dawkins, a leading popularizer of the view that evolutionary theory has removed any need for God, Pope Benedict's statement must seem downright ludicrous. Contrary to Benedict, Dawkins views religious faith as a crutch that obstructs human reason rather than a tool that can unleash it. He argues that "faith is the great cop-out, the great excuse to evade the need to think and evaluate evidence. Faith is the belief in spite of, even perhaps because of, the lack of evidence."[5]

However, Dawkins' caricature of faith bears no resemblance to the Catholic understanding of the term. From a Catholic perspective, faith is seen not as the antithesis of reason, but rather as a necessary partner with reason in the ongoing search for the ultimate truth. For Catholics, as John Paul II stated, "faith asks that its object be understood with the help of reason"[6] rather than in spite of reason. In light of this, Dawkins' view of faith is the *exact opposite* of the Catholic understanding of the term. He confuses faith with fideism, the belief that understanding comes from faith alone, often at the expense of (or in opposition to) the use of human reason, which is a theological position distinctly at odds with the Catholic faith.

Dawkins fails to capture not only the Catholic understanding of faith but even the popular everyday use of the term. For example, when a man states that he has faith in his wife, he normally states this because he has *evidence* to justify that faith—his wife has given him *reason* to have faith in her based upon her actions and words. Faith for the husband is not a blind naïve trust that flies in the face of the evidence. Rather, it flows from the abundant amount of evidence provided by countless instances of his wife's loving and caring behavior. Faith in this

5. Richard Dawkins, Speech at the Edinburgh International Science Festival, April 15, 1992; reprinted as "EDITORIAL: A Scientist's Case against God," *The Independent*, 1992, p. 17.

6. John Paul II, *Fides et Ratio* 42, encyclical letter, September 14, 1998, vatican.va.

context is usually consistent with what is known through the use of human reason.

While faith can certainly be reasonable, it is important to recognize that faith is not scientific in nature. This is an important distinction for the evolution/creation discussion, as many like Dawkins erroneously equate reason with science. The case of the husband described above can illustrate aptly this distinction between reason and science. While it would be quite reasonable for the husband to have faith in his wife if she has exhibited years of loving behavior toward him, there is no scientific experiment that one can perform to measure, quantitate, or justify the amount of faith the man should have (or actually does have) in his wife. Of course, one could scientifically study the wife's behavior and attempt to quantify the number of specific "loving" actions or behaviors she exhibits. However, such a study could never justify the man's faith *scientifically* because it could never get at his wife's inherent motivations, her beliefs, or her actual love for her husband, as these are not physical things that are amenable to scientific investigation. Science, at least modern empirical science as currently understood, is limited to observing and measuring her physical behavior, whether this be the physical actions she performs or the physical changes that occur in her brain. It takes a rational human being to interpret these actions and changes as loving, as coming from a person in whom it is worthy to place one's faith. While such an interpretation is not scientific, it can certainly be rational and reasonable.

In a similar fashion, when one says that he or she has faith in God's existence, one is not merely taking this position in spite of the evidence. Rather, faith in God's existence is a rational position for which one can offer well-reasoned arguments. As the *Catechism of the Catholic Church* points out, "The Church holds and teaches that God, the first principle and last end of all things, can be known with certainty from the created world by the natural light of human reason."[7]

Again, it is important to realize that these demonstrations, though reasonable, are not scientific in nature. Rather, the arguments for the existence of a Creator, some of which date back to the ancient pagan philosophers, are philosophical in nature. Such arguments have a rich history within the Catholic tradition and include St. Anselm's ontological argument and St. Thomas Aquinas' five ways. Aquinas' first way illustrates the nature of these types of rational but non-scientific arguments. In his first way, Aquinas begins with the claim, which he justifies based on human experience, that anything in motion must have been set in motion by another object. Every time we observe an object set in motion, this motion or change is triggered by another object, *not* the object that undergoes the change in motion itself. However, this first object, which triggers the motion of the second, must have been itself set in motion/change by another object, and so forth. Aquinas argues that this series of causes cannot go on *ad infinitum*, as there would be nothing to generate the first movement. He then concludes, "Therefore it is necessary to arrive at a first mover, moved by no other; and this everyone understands to be God."[8]

While much has been written regarding the validity of Aquinas' first way, there is no experimental body of scientific work that can address this argument. Science can measure how physical objects move or how momentum is transferred from one physical object to the next, but there is no scientific experiment one can perform to answer the question of why movement exists in the first place. One can certainly quibble, as numerous authors have, with the logic Aquinas employs to build his argument, but that is beside the point here. The key is to recognize that one

7. *Catechism of the Catholic Church* 36.

8. Thomas Aquinas, *Summa theologiae* 1.2.3. All quotations from the *Summa* are from second and revised edition, 1920, trans. Fathers of the English Dominican Province, from newadvent.org.

can make rational arguments about faith and God and that such arguments are not scientific in nature. Rather, such arguments 1) are philosophical and/or theological in nature and 2) can provide useful knowledge about reality. Recognizing that there is real knowledge to be gained outside of the narrow confines of modern science, that science doesn't hold all the answers, is what allows Ratzinger to maintain that evolution and creation are "two complementary—rather than mutually exclusive—realities," and it is an essential first step toward integrating Catholicism and the science of evolution.

LIMITS OF SCIENCE

In our increasingly technologically reliant society, scientific knowledge holds a privileged place. Scientific departments proliferate at modern universities while departments devoted to theology and philosophy continue to be downsized and struggle with enrollment. On the one hand, science seems to provide knowledge that can improve our health and quality of life, while philosophy and theology seem to only produce unresolvable squabbles. Given this perception, it is little wonder that scientism, the belief that science is the only path toward real knowledge, continues to gain adherents. The advocates of scientism operate under the assumption that the material world is all that exists and that modern science alone has the resources to answer all the meaningful questions.

The adherents of this position tend to frame themselves as the guardians of rational inquiry, defending it against the antiquated superstitions of theology and the unfounded speculations of philosophy. But at its heart, scientism has a fatal flaw, one articulated succinctly by the philosopher Ed Feser: "Despite its adherents' pose of rationality, scientism has a serious problem: it is either self-refuting or trivial. . . . The claim that scientism is true is not itself a scientific claim, not something that can be established using scientific methods. Indeed, that science is even a rational form of inquiry (let alone the only rational form of inquiry) is not something that can be established scientifically."[9]

Feser makes the point that scientism is a philosophical position, and as such, it can only be defended by philosophical arguments. But for an adherent of scientism, philosophical arguments hold no value. This leaves the adherent of scientism in the unenviable situation of advocating for a position that he has no capability of defending given the limited tools he has left at his disposal.

In addition to being trapped inside a self-refuting philosophy, the practitioners of scientism make matters worse by seeking to answer questions that the scientific method has no ability to handle. An excellent example of this overreach can be found at the beginning of the biologist Richard Dawkins' book *The God Delusion*, where he confidently states that "the existence of God is a scientific hypothesis." However, as Dawkins attempts to build his "scientific" case against God, he unwittingly illustrates both the futility of such an exercise and the futility of scientism in general.

To construct his case, Dawkins invokes the results of what he calls the great prayer experiment, a scientific experiment in which researchers examined the outcomes of two groups of sick patients, those who were receiving intercessory prayer and those who were not. The study found that there was no difference in measurable health outcomes between the two groups, a result that Dawkins believes provides clear scientific evidence for the nonexistence of God. The results of the experiment, however, do nothing of the sort. The experiment merely provides evidence that 1) God is not what Dawkins thinks he is, an easily manipulated accountant that adds up prayers and doles out rewards based upon some predetermined calculus, and 2) prayer is not what Dawkins thinks

9. Edward Feser, "Blinded by Scientism," *Public Discourse*, March 9, 2010, https://www.thepublicdiscourse.com/2010/03/1174/.

it is, a human petition whereby people manipulate a higher power to achieve a desired physical outcome.

Because Dawkins believes everything must be reduced to a materialistic explanation, he treats God and prayer in the same manner as he would treat physical entities such as a medication or a surgical treatment. However, if prayer is not merely a physical thing (and it isn't), then it is not the type of entity that is reducible to scientific inquiry.

This is not to claim that prayer doesn't have physical manifestations; it clearly does. The act of praying can be correlated with physical changes such as heart rate reductions or alterations in neural firing patterns. But, for the Christian, the act of prayer transcends the physical because at its essence, prayer is communion with God. As such, while the physical correlates of prayer may be accessible to scientific study, the true essence of prayer remains beyond the purview of science. For a Christian, prayer is the opening of oneself up to the will of the Creator, a divine will that no scientific experiment can measure. It is asking God that "thy will be done." This stands in stark contrast to Dawkins' view of prayer as an attempt to bend the will of the Creator to match our desires.

From a Christian perspective, both answered and unanswered prayers are wholly consistent with an omnipotent Creator who has a better grasp on our true needs than we do. Any possible outcome of the great prayer experiment is entirely consistent with the existence of the Christian God (not to mention other versions of God, such as deistic versions). The results of the experiment have no bearing on the question Dawkins is asking: the existence or nonexistence of the Christian God.

Dawkins, like many others who are immersed in scientism, is trying to use science for something it is not equipped to do. To his credit, he is correct that the existence or nonexistence of God is worth investigating, but the way he sets about doing so is bound for failure. As the Protestant theologian Keith Ward points out, "The existence of God is certainly a factual question [God either exists or he doesn't], but it is not a scientific one."[10]

Recognizing the limits of science and what is and what is not a scientific question gets at the heart of the evolution and creation discussion. In fact, not acknowledging the limits of science is one of the biggest obstacles that hinders a more fruitful engagement between the science of evolution and a Catholic understanding of creation. The recognition that modern science has limits, that it provides rational explanations regarding only how the *physical* world is structured and operates, is essential for any effort to successfully integrate evolution and creation.

Science is not the sum of all our knowledge. While science is an extremely powerful and useful tool for addressing how the physical universe operates, humans naturally do not end their inquiries there. It is quite normal for humans to inquire into not only how the universe is structured but also *why* it is structured in the manner described by science. Humans quite naturally inquire regarding the ultimate purpose and meaning of things. All great literature and art touch on these big questions regarding the purpose of existence, the purpose of human life, or the purpose of the universe. In a similar manner, when one examines the process of evolution from a scientific perspective, it is quite natural for this scientific investigation to spur deeper, more penetrating questions regarding the ultimate purpose of evolution. Does evolution operate to fulfill God's plan and purpose? Is evolution devoid of any sense of purpose or meaning? Do humans have any more worth or value than other primates?

All of these questions are very real and very relevant to how we live our lives. However, they are not scientific questions because the subject matter, ultimate purpose, is not a physical entity that is quantifiable in any meaningful scientific way. Purpose is something that lies outside this realm; it is a question beyond

10. Keith Ward, *The Big Questions in Science and Religion* (West Conshohocken, PA: Templeton, 2008), 30.

the ontological limits of science. Attempting to use science to demonstrate that humans have no purpose or meaning is akin to using a hammer to try and fix your desktop computer when it doesn't boot up properly. It is the wrong tool for the job.

FIDEISM AND YOUNG EARTH CREATIONISTS

While advocates of scientism sever the connection between faith and reason by making science the sole arbiter of rational thought, there is an equal but opposite problem that exists on the flip side of the evolution/creation spectrum. Young Earth Creationists (YECs), rather than looking to evolutionary science to provide insights into legitimate scientific questions, rely upon the revelation of Scripture as the ultimate arbiter of natural history.[11] They rely upon the historical accuracy of the scriptural account even if it is not consistent with the scientific data garnered through the proper application of human reason. In this context, faith is seen as something that can and should override scientific reasoning if that reasoning legitimately comes to a conclusion that does not coincide with some narrow understanding of the faith.

Of course, the Church recognizes that human reason is prone to mistakes and that in a fallen world, greed, pride, and arrogance can corrupt human reason such that it distorts rather than enlightens our understanding. Pope Pius XII made this point in the 1950 encyclical *Humani Generis*: "There are not a few obstacles to prevent reason from making efficient and fruitful use of its natural ability. . . . Hence men easily persuade themselves in such matters that what they do not wish to believe is false or at least doubtful."[12] But despite this concern, Pope Pius XII makes it clear that human reason, on its own, *is* capable of coming to know God: "Human reason by its own natural force and light can arrive

11. The term Young Earth Creationists here refers to those who regard the Genesis text as a historical description of the events of creation and believe that the earth is young, less than ten thousand years old.

12. Pius XII, *Humani Generis* 2, encyclical letter, August 12, 1950, vatican.va.

at a true and certain knowledge of the one personal God, Who by His providence watches over and governs the world, and also of the natural law, which the Creator has written in our hearts."[13]

Pope Pius XII's confidence that human reason can bring one to knowledge of God flows from the Catholic understanding that the human capacity for reason reflects God, the ultimate source of reason. The Creator God has endowed humans with the rational capacity to come to knowledge of him and his rationally ordered creation through the application of our reason. Therefore, human reason, when properly applied, either philosophically or scientifically, can draw us closer to the truth about both God and his creation.

This appreciation for the capability of human reason is a stumbling block for many YECs who see faith alone, particularly as revealed through a historical reading of Scripture, as the sole means to brings us to the truth. This is a position that naturally leads to a denigration of the capacity of human reason, for if one takes on faith that God created the world in six days (as described by the author of Genesis), one must then put aside what human reason has uncovered regarding the age of Earth, the age of the universe, or the evolutionary connections between living organisms. This is exactly the position in which staunch YECs find themselves. Hugh Owen, a Catholic YEC, illustrates this type of reasoning: "Since it is impossible to reconcile the long ages of evolution with the traditional doctrine of creation as set forth, for example, in the *Roman Catechism*, without 'recasting' it, this . . . effectively rules out any possibility that theistic evolution could be deemed acceptable for Catholics by the Magisterium."[14]

Another Catholic YEC, Fr. Peter Fehlner, displays a similar rationale for rejecting the scientific data indicating that the Earth is old: "The Creator, being the only witness to what happened

13. *Humani Generis* 2.

14. Hugh Owen, "Hugh Owen's Response to Cardinal Ruffini on the Days of Creation," https://www.pcpbooks.net/hugh_owen_and_cardinal_ruffini.html.

in the beginning, has told us He made the stars and made them shine within a period of 24 hours, thus providing a key to the interpretation of the appearances 'in the beginning.'"[15] Thus, even if the preponderance of the scientific evidence indicates that the Earth is 4.5 billion years old, one is required to explain away or ignore this data in order to maintain what is "known" by faith.

Ironically, this creates a situation in which both extremes in the evolution debate, the YECs and the atheistic evolutionists like Richard Dawkins, have become strange bedfellows. Both groups have deliberately severed the connection between faith and reason (although they have cut the cord from opposite ends). As John Paul II articulated in *Fides et Ratio*, "Faith and reason are like two wings on which the human spirit rises to the contemplation of truth."[16] It is not surprising then that the severing of this crucial link can lead both groups astray.

Once this cord is cut, both the atheistic evolutionists and the YECs tend to place God in competition with other possible explanations for the emergence of organisms. As the philosopher Conor Cunningham has pointed out, "The only difference is that the [atheistic evolutionists] think [this God] does not exist, or that at most it is a blind watchmaker."[17] For YECs, the commitment that God crafted the universe in the manner described in Genesis inherently puts their view of God in conflict with any scientific explanations regarding the evolution of the universe. They have—unwittingly perhaps—set up a situation in which God is forced to compete with creaturely causes to maintain his relevance: either God is the cause, or some scientific process is the cause. From this perspective, the establishment of a robust scientific account regarding the emergence of stars, planets, and organisms would necessarily remove God from the picture.

15. Peter D. Fehlner, "In the Beginning," Kolbe Center for the Study of Creation, November 1, 2001, https://www.kolbecenter.org/in-the-beginning/.

16. *Fides et Ratio*.

17. Conor Cunningham, *Darwin's Pious Idea: Why the Ultra-Darwinists and Creationists Both Get It Wrong* (Grand Rapids, MI: Eerdmans, 2010), 151.

This, though, is the exact same limited framework from which atheistic evolutionists operate. They see evolutionary explanations for the emergence of organisms as replacing the need for God. God is seen as one cause among many, and a relatively impotent one at that. Science, on the other hand, is seen as offering a more accurate explanation of cosmic and biological evolution. It is for this reason that Dawkins claimed the science of evolution allowed him to be an intellectually fulfilled atheist. If the science of evolution can explain everything, then Dawkins can echo Laplace's famous remark regarding the need to invoke God: "I have no need for that hypothesis."

Yet the God of the YECs and the atheistic evolutionists is not the God that Catholics worship. The Catholic Creator God is not one cause among many, but rather the source of all existence. He is the reason that anything exists at all. He is the reason that stars can burn and produce heavier elements, and he is the reason that species can become modified and shaped by their interactions with the environment. As the *Catechism* points out, the Church's understanding of creation is meant to address "the basic question that men of all times have asked themselves: 'Where do we come from?' 'Where does everything that exists come from and where is it going?'"[18] It is meant to address the question of why things exist in the first place.[19]

While these questions will be expanded upon throughout this book, the key point to grasp up front is that for Catholics, creation is not a point in time or some isolated act. Rather, it is a deep metaphysical reality. It is the continual reliance of everything and everyone on God for existence at every point in time. From this metaphysical perspective, creation is neither a six-day process nor a fourteen-billion-year process (although the question of how God's creative act has unfolded is a related and

18. *CCC* 282.

19. Certainly, the questions of how the material world came into existence or how humans came into existence are related to this question, but those questions are secondary to the question of existence itself.

interesting question to be explored in this book). Creation is the utter dependence of everything upon the primary causality of God for its existence. As the *Catechism* describes it, God's creative power "not only gives [his creatures] being and existence, but also, and at every moment, upholds and sustains them in being, enables them to act and brings them to their final end."[20]

Thus, physical creatures and physical forces only operate and produce evolutionary changes because God as the primary cause of all that exists "enables them to act." From the Catholic perspective, God is not competing with natural causes but is the ultimate source of such causes. The evolutionary processes that shape new species can only do so in union with the God who endows and sustains such processes with both existence and efficacy. Such a view rightly places God on a different plane than creaturely causes and frees up the intellectual space to see evolution and creation as "two complementary realities,"[21] rather than mutually exclusive realities in competition with one another.

EVOLUTION AND CREATION: THE INTEGRATION OF SCIENCE AND FAITH

To properly view evolution and creation as complementary rather than competing realities, it is essential to adequately frame the relationship between science and faith. There are many different models regarding how the knowledge gained through faith is related (or not related, as seen in the previous section) to the knowledge gained via science. One popular model was put forth by the late Harvard paleontologist Stephen J. Gould. Gould advocated for the non-overlapping magisterial (NOMA) view of science and faith, a view he believed was necessary to eliminate any conflict between the two.

He argued that the two disciplines were entirely distinct

20. *CCC* 301.

21. Ratzinger, *"In the Beginning,"* 50.

from each other such that knowledge gained in one discipline had no bearing on the other. Because science has a distinct set of questions it addresses and theology/faith address an entirely different set of questions, they each can happily occupy their own well-defined arenas. If each discipline were to stay within its respective boundaries, opportunity for conflict would be minimized. Upon first examination, the NOMA idea seems both reasonable and helpful. For example, it is certainly important to recognize that science is not the tool to investigate the nature of God nor is theology the tool to investigate the physical processes of evolution.

But this approach has limits because the questions that we ask about the world are not always neatly sequestered into specific disciplinary categories. In fact, inherent in the most profound questions that humans ask—What is the meaning of life? What is the purpose of the universe?—is a desire for gaining a unified vision of the whole rather than settling for a fragmented piecemeal picture. As every discipline has specific methodological limits, no one discipline, operating in isolation, can provide a complete picture of reality. Just as a symphony requires that each instrument be properly tuned and integrated into the whole to produce a beautiful concerto, each discipline needs to be tuned in its own proper methods and integrated with the knowledge gained from other disciplines to provide the most complete picture of reality. If one instrument overpowers the others, the product is a distorted, unbalanced cacophony.

A distorted, unbalanced cacophony is also what happens if science, philosophy, and theology do not interact properly. Because many of the questions raised by evolutionary theory deal with origins—the origin of life, the origin of man—science, theology, and philosophy all must play a role in addressing these. For example, a complete picture of man requires an understanding of both what man is and how man's physical form came to be. As the philosopher Mortimer Adler pointed out, the question of man is "a mixed question, a question that cannot be adequately answered either by scientists alone or by philosophers alone, but only by their collaboration—by combining the findings of scientific investigation with the contributions of philosophical analysis and criticism."[22] The trick is getting these disciplines to interact in a productive fashion. Pope Pius XII recognized this need in *Humani Generis* when he stated that it is critical for the implications of evolution to be investigated "on the part of men experienced in both fields [human sciences and sacred theology]."[23]

How then is this done in practice, particularly in the case of understanding the origin of man? For starters, one must recognize that the question of man's origins cannot be fully separated from the question of what man is. From a Catholic perspective, man is a unity of body and soul, a unity that reflects the fact that man is made in the *imago dei*. While man is a physical creature, man is also endowed with an immaterial spiritual soul, which is considered the animating principle of the body. This unity of body and spiritual soul gives rise to "a being at once corporeal and spiritual,"[24] a being that the Church teaches is not reducible to the material. This theological position, though, is not merely taken on faith, as there are good philosophical arguments to support it.

In fact, many philosophers, even agnostic ones, have argued that the abilities of the human intellect represent capabilities that are not fully reducible to the material.[25] The ability to grasp universal concepts that have no correlates in the material world, the subjective perception of qualia such as "blueness" or "sweetness" that seem to lack a physical nature, the perception of universal moral laws, and the ability to perform conceptual and symbolic

22. Mortimer Adler, *The Difference of Man and the Difference It Makes* (New York: Fordham University Press, 1993), 13.

23. *Humani Generis* 36.

24. *CCC* 362.

25. Edward Feser's *The Philosophy of Mind: A Short Introduction* is an excellent foray into this area.

thought are just a few examples that suggest human rational capabilities are not fully reducible to physical phenomena. It is not necessary to delve into the validity of such arguments here, as that goes well beyond the scope of the book. For present purposes, it is sufficient to recognize that such philosophical arguments exist and that 1) rigorous "philosophical analysis and criticism" is necessary to resolve them and 2) their resolution influences the role science plays in understanding the origin of man.

For example, if one comes to the philosophical conclusion that all of man's abilities can be reduced to physical phenomena, then there would be an expectation that science could explain the totality of man's origins, including the origin of his mental abilities and moral sensibilities. Likewise, if rigorous philosophical analysis leads one to the conclusion that man has an immaterial aspect to his nature, such a conclusion would limit what one would expect science to contribute to the question of human origins.

The Church has a deep respect for science and its ability to expand human understanding. At the same time, though, the Church realizes that science does not hold exclusive rights to knowledge and that scientific knowledge, if it is to truly serve humanity, must be integrated into a wider whole. The science behind man's moral sensibilities is a case in point. There is certainly a material aspect to our moral sensibilities that science can investigate. It is known via science that feelings of guilt, empathy, and anger at injustice are all associated with changes in brain chemistry, brain chemistry that is shared with other primates. Furthermore, based upon one's brain chemistry, it is known that certain people struggle with guilt or anger or a lack of empathy more than others. Understanding this scientific aspect of moral sensibilities can enlighten our understanding of the theological implications of human actions and the moral culpability of human actors. As John Paul II stated, "Science can purify religion from error or superstition."[26] But what science cannot do is explain adequately our transcendent moral sensibilities in the first place. While certain expressions of our moral sensibilities may be rooted in our animal physiology, our knowledge of right and wrong transcends such emotions.[27]

Likewise, while science has been tremendously successful in acquiring knowledge of the material world, there are many questions that it is simply ill-equipped to address. For example, what is the ultimate purpose of human existence? Or why does something exist rather than nothing? Or why has our human reason been so successful in understanding the natural world? Science cannot adequately address these transcendent philosophical questions given its specific methods. However, the knowledge science has gained regarding the material world is useful for properly framing such questions. For example, if the scientific data indicates that man's body has evolved via physical evolutionary processes from other primates, then the proper philosophical question to ask is whether these physical processes can explain all there is to know about mankind. As outlined above, Catholics have sound theological and philosophical reasons for believing that the answer to this question is no: man is more than a material being. This recognition flows from the integration of two truths: the scientific truth that man has evolved and the religious truth that man was created in the *imago dei*.

26. John Paul II, "Letter of His Holiness John Paul II to Reverend George V. Coyne, SJ, Director of the Vatican Observatory," June 1, 1988, vatican.va.

27. In fact, trying to reduce our moral and ethical sense to the product of a material evolutionary process actually eradicates the basis for any moral or ethical foundation. If our moral and ethical sense is nothing more than a survival mechanism cobbled together by an irrational and amoral material process, there is nothing left to ground human ethics other than the arbitrary choices of those who hold power. We are left with no objective right or wrong, having nothing more than what evolution has conditioned us to think in order to survive. This strips away any solid foundation for morality or ethics, a foundation that is needed to build flourishing human societies. As Pope Benedict has stated in *Truth and Tolerance* (San Francisco: Ignatius, 2004), "Here, the attempt to distill rationality out of what is in itself irrational quite visibly fails."

CATHOLIC EVOLUTION; CATHOLIC CREATION

Over the past few decades, the Church has pursued multiple efforts to properly integrate the science of evolution with a Catholic understanding of creation. Under the pontificate of John Paul II, the Vatican Observatory sponsored a conference and published the proceedings as a book entitled *Evolutionary and Molecular Biology: Scientific Perspective on Divine Action.* John Paul's successor, Pope Benedict XVI, was similarly interested in the topic. Prior to his election as pope, a Vatican symposium on the topic of "Evolution and Christianity" was sponsored by the Congregation for the Doctrine of the Faith while he served as its prefect. In addition, after becoming pope, Benedict held a 2005 symposium with many of his former doctoral students on the topic of "Creation and Evolution."

In fact, papal statements and Church documents, in recent years, have been particularly open to the scientific knowledge gained through the study of evolution. In 1996, John Paul II remarked that "new knowledge leads to the recognition of the theory of evolution as more than a hypothesis."[28] Even the current version of the *Catechism* speaks in respectful terms regarding evolution: "The question about the origins of the world and of man has been the object of many scientific studies which have splendidly enriched our knowledge of the age and dimensions of the cosmos, *the development of life-forms* and the appearance of man."[29] While the *Catechism* does not delve into the specifics, it seems apparent that the scientific studies referred to by the *Catechism*, studies that have "splendidly enriched our knowledge" regarding "the development of life-forms," involve the process of evolution.

Now, it is important to recognize that these statements do

28. John Paul II, "On Evolution" 4, Message to the Pontifical Academy of Sciences, October 22, 1996, in *Origins* 26, no. 25 (December 5, 1996): 415.

29. *CCC* 283 (emphasis added).

not indicate that the Church endorses the science of evolution (as mentioned previously, the Church does not endorse scientific theories), nor do they indicate that one must agree that life forms evolved to be a Catholic in good standing. The positive mention of the science of evolution in the *Catechism* (or in an address, a homily, or a papal general audience) holds quite a different magisterial weight than the Church's teaching in the *Catechism* and elsewhere on the dogma of the Trinity or the Real Presence of Christ in the Eucharist.[30]

Despite the lower magisterial weight, the faithful can still profit from what popes have stated regarding evolution in these documents or addresses. For example, John Paul II maintained that evolution, if properly understood, does not take precedence over creation, as so many atheistic evolutionists argue. In fact, he correctly points out that evolution is subordinate to creation: "Rightly comprehended, faith in creation or a correctly understood teaching of evolution does not create obstacles: Evolution in fact presupposes creation."[31]

The utter dependence of evolution on creation gets to the heart of the Catholic approach to the matter. To fully comprehend this relationship, though, it is essential that the terms creation and evolution are properly understood. If creation refers to a six-day period during which the heavens and earth were fashioned, it is nonsensical to say evolution is dependent upon creation. A similar problem is encountered if evolution refers to a materialistic process that can explain the entirety of man. However, if creation refers to the reliance of all things on God for their existence and evolution refers to a scientific explanation for physical phenomenon, one then has an adequate framework from which to operate.

30. A short summary of the differences in magisterial weight given to different documents as well as different statements within these documents is presented in chapter 3.

31. John Paul II, quoted in Christoph Schönborn, "Creation and Evolution: To the Debate as It Stands," *Zenit*, October 2, 2005, https://www.ewtn.com/catholicism/library/cardinal-schnborn-on-creation-and-evolution-10246.

Properly defining and using the terms creation and evolution is key. That is why it is essential to recognize that the Catholic understanding of creation is *not* synonymous with a historical reading of the six-day creation account in Genesis. There is not now, nor has there ever been, a requirement that Catholics read the first few chapters of Genesis as a historical description of events. In fact, as will be discussed in later chapters, many of the Church Fathers did not read the first chapters of Genesis in this manner. Furthermore, the Church specifically rejects biblical fundamentalism, the position that the entire Bible corresponds exactly to historical fact. The Church realizes that the divinely inspired biblical authors frequently used figurative or poetic language to make salient points.

Equally important to remember, though, is that the term evolution is *not* synonymous with atheism or materialism, despite what many atheistic evolutionists would have people believe. In fact, evolution, as a purely scientific theory, cannot be used to demonstrate the philosophical position that God does not exist or that the material world is all that exists. While the science of evolution can certainly be employed by atheists in support of their philosophical arguments, it can also be employed to good effect by theists to support their own philosophical arguments regarding God. For example, the biologist Nicanor Austriaco, OP, echoing an argument from Aquinas regarding the grandeur of God being reflected in the many and diverse creatures that exist, used evolution to make the following philosophical claim: "Therefore, in my view, it was also fitting that God created via evolution rather than via special creation because in doing so he was able to create more species to reflect his glory: Four billion species created over a three-billion year [plus] period rather than just the eight million extant species today."[32]

This type of positive engagement with evolutionary thought

32. Nicanor Austriaco et al., *Thomistic Evolution: A Catholic Approach to Understanding Evolution in the Light of Faith* (Providence, RI: Cluny Media, 2019), 147.

is echoed in the writings of Pope Benedict XVI and Pope St. John Paul II. Neither seemed overly concerned that the *scientific* knowledge surrounding evolution would undermine the core tenets of the Catholic faith. While they cautioned against atheistic philosophical systems that have been speciously linked with evolutionary ideas, they were confident that the Church had nothing to fear from the *science* of evolution. This confidence stems from an understanding, articulated by John Paul II in his encyclical *Fides et Ratio*, that "the unity of truth is a fundamental premise of human reasoning."[33] In other words, he is confident that the truth found through scientific research will not conflict with the truths of the faith because all truth flows from the same source, the Creator God. Therefore, as the *Catechism* articulates, "methodical research in all branches of knowledge, provided it is carried out in a truly scientific manner and does not override moral laws, can never conflict with the faith, because the things of the world and the things of the faith derive from the same God."[34]

All roads of human inquiry then, whether they be biological, psychological, theological, or philosophical in nature, if properly followed, should converge upon the ultimate source of truth, the one triune God. Given this understanding of the unity of knowledge, one can be assured that any apparent conflicts are just that: apparent. They arise as the result of either an improper understanding of the science or an incomplete understanding of the faith.

But even for those who are fixated upon the irreconcilability they perceive between the Catholic faith and the science of evolution, the exquisite order and beauty that the science reveals holds within it the possibility of reconciliation. This beauty is evident even to atheists like Richard Dawkins, who stated that the science of evolution reveals "a sinewy elegance, a poetic beauty that

33. *Fides et Ratio* 34.
34. *CCC* 159.

outclasses even the most haunting of the world's origin myths."[35] Even Dawkins sees the unmistakable beauty and order in nature, an observation that can naturally cause one to ask the deeper question: *Why* is the universe so beautifully fit and ordered for the evolution of life?

While there are a variety of possible answers to this question, they all eventually boil down to two distinct possibilities. The first possibility is to declare that the universe just merely is this way, that the universe and its underlying structure is a brute fact that cannot be explained nor needs to be explained. It just is. This position was popularized by the twentieth-century analytic philosopher Bertrand Russell, who stated, "I should say that the universe is just there, and that's all."[36]

Many, though, find this option intellectually unsatisfying for obvious reasons. For those who do, the alternative possibility is to declare that the universe is the product of the mind of God, the Creative Reason who gives existence and order to all things. Characteristically, Pope Benedict XVI distinguishes these two possibilities quite elegantly: "Ultimately it comes down to the alternative: What came first? Creative Reason, the Creator Spirit who makes all things and gives them growth, or Unreason, which, lacking any meaning, strangely enough brings forth a mathematically ordered cosmos, as well as man and his reason. . . . We believe that at the beginning of everything is the eternal Word, with Reason and not Unreason."[37]

Such a belief in the Creative Reason is not antiscientific, nor is it unreasonable. The belief that the Creative Reason brought the universe into existence gives us warrant to believe that the universe is rational and that we may approach an understanding of not only the order in the created universe but also the ultimate

35. Richard Dawkins, *River Out of Eden* (New York: Basic Books, 1995), preface.

36. Bertrand Russell and Frederick Copleston, "Debate on the Existence of God,"1948, reprinted in *The Existence of God*, ed. John Hick (New York: Macmillan, 1964), 175.

37. Benedict XVI, *Creation and Evolution*, 214.

source of that order. It is this belief that assures Catholics that through the proper exercise of human reason, inspired by divine reason, one can reach the conclusion that evolution and creation are "two complementary—rather than mutually exclusive—realities."[38] To unpack what Pope Benedict XVI meant by this, it is first worth examining the Church's historical engagement with evolutionary theory.

38. Ratzinger, *"In the Beginning,"* 50.

3

The Church's Historical Engagement with Evolutionary Theory

> Today, nearly half a century after the publication of [*Humani Generis*], new knowledge leads to the recognition of the theory of evolution as more than a hypothesis. It is indeed remarkable that this theory has been progressively accepted by researchers following a series of discoveries in various fields of knowledge. The convergence, neither sought nor provoked, of the results of work that was conducted independently is in itself a significant argument in favor of this theory.[1]
>
> —Pope St. John Paul II

There is a common misperception, fueled largely by the lingering after-effects of the Galileo affair, that the Catholic Church has been involved in a long-standing battle with science. According to this view, the Catholic Church has been waging a continual yet ultimately fruitless counter-offensive against the inevitable march of scientific progress. This belief is so common that in *Galileo Goes to Jail and Other Myths about Science and Religion*, a popular book authored by a group of preeminent historical scholars, an entire chapter was dedicated to debunking the myth that the medieval Church suppressed the growth of science.[2]

While this myth tends to maintain a popular hold on the

1. John Paul II, "On Evolution" 4, Message to the Pontifical Academy of Sciences, October 22, 1996, in *Origins* 26, no. 25 (December 5, 1996): 415.

2. Michael Shank, "That the Medieval Christian Church Suppressed the Growth of Science," in *Galileo Goes to Jail and Other Myths about Science and Religion*, ed. Ronald L. Numbers (Cambridge, MA: Harvard University Press, 2009), 19–27.

public imagination, for those willing to look at the actual history of the Church's engagement with science, it becomes readily apparent that the myth cannot stand up to scrutiny. From the Middle Ages to the present, the Catholic Church has been a major financial supporter of scientific research, and many Catholics were and continue to be found among the cutting-edge scientists of the day. In addition, Catholic clergy, from Copernicus to Mendel to Lemaître, have found themselves at the forefront of scientific discovery in every era.

Given the Church's view that scientific discovery is a means of understanding the Creator, the abundance of Catholic scientists down through the ages should not come as a surprise. In fact, in the tradition of the Church, there are two great books through which God has revealed himself—the Book of Scripture and the Book of Nature. As the Scripture proclaims, "For all people who were ignorant of God were foolish by nature; and they were unable from the good things that are seen to know the one who exists, nor did they recognize the artisan while paying heed to his works" (Wis. 13:1). From a Catholic perspective, an understanding of creation generated via the scientific pursuit can foster a deeper understanding and reverence of God as Creator. As a result, the Church has always had theological reasons to support the advancement of science.

The Church's support was particularly evident during the Middle Ages when it was *the* major benefactor of science. Certainly, there were practical reasons for the Church to support science during this period, such as the need for more accurate calendars, better navigation, etc., but these reasons were augmented in a particular manner by the Church's understanding of the connection between creation and Creator. Taken together, it is not surprising that the Church was instrumental in establishing and maintaining the university system in Europe, a system that fostered both rational inquiry and scientific discovery. Most impartial historians have noted that the Church was the one institution in Europe at the time that was interested in both the preservation of knowledge and the pursuit of new knowledge through the application of human reason. It was a time that saw Catholics, many of whom were clergy, make major advances in the fields of geology, astronomy, and biology.

Given the fertile breeding ground for scientific inquiry within the Catholic Church of the Middle Ages, many of the leading figures of the subsequent Scientific Revolution (sixteenth and seventeenth centuries) were believing and practicing Catholics.[3] Catholic scientists included towering figures such as Copernicus, Galileo, and Pascal, men whose groundbreaking work advanced the fields of mathematics, physics, and astronomy. In addition, the Church maintained some of the best astronomical observatories in Europe during this period, and it continued to support financially many branches of scientific research. This congenial relationship continues to this day with the maintenance of the Vatican observatory, the support of scientific research conferences, and the regular consultations that popes have had with the leading scientific thinkers of the day on issues such as evolution and neurobiology.

Despite the largely agreeable relationship between science and the Church over the centuries, there are those who still point to the Galileo affair as evidence that the Church harbors a latent hostility to science. Yet the Galileo case seems to be the exception that proves the rule. One cannot point to a single other case throughout history where a scientist was condemned for and had to publicly recant his *scientific* position. It stands as such an anomaly that centuries later John Paul II felt compelled to issue an apology for the Church's role in the affair.

Before examining the Church's interaction with evolutionary theory, it is worth touching on a few points regarding the

3. For a good overview of the number of practicing Catholic scientists who played major roles in the Scientific Revolution, see Nicholas Spencer's *Magisteria: The Entangled Histories of Science and Religion* (London: Oneworld Publications, 2023).

Galileo affair given its historical importance in shaping popular perceptions of the Church/science relationship.[4] In particular, the Galileo case demonstrates that supposed conflicts between religion and science are always much more complex than they are commonly made out to be.

WHAT REALLY HAPPENED TO GALILEO

Galileo lived during the late 1500s and early 1600s in the immediate aftermath of the Protestant Reformation, a historical circumstance that certainly influenced the Church's response to Galileo's writings. Based upon the prior work of Copernicus as well as his own scientific observations, Galileo advocated for a heliocentric model of the heavens, one in which the Earth revolved around the sun. While Galileo amassed a good deal of evidence for heliocentrism, a key piece of evidence, the observation of parallax, was missing. If the Earth was indeed moving around the sun, observers should be able to detect parallax, shifts in the relative position of stars in the night sky depending on where the Earth was in its orbit. Due to the lack of adequate equipment and the fact that stars were much farther from the Earth than people at the time thought, parallax was not detected until long after Galileo's death.

Without this evidence, some astronomers were skeptical of the heliocentric theory and advocated for versions of the more traditional geocentric system, which held that the Earth was stationary at the center of the cosmos. The Church had other reasons for concern regarding Galileo's heliocentric ideas. The first and most important had to do with the interpretation of Scripture. There were several biblical passages—Joshua 10:13, Ecclesiastes 1:5, and Psalms 93 and 104—which, if taken literally, indicated that the Earth was stationary. If the heliocentric theory

4. Readers who are interested in an in-depth look at the Galileo affair can find a well-written account in *Galileo Revisited* by Paschal Scotti (San Francisco: Ignatius, 2017).

was correct, how were these to be interpreted and who had the authority to decide upon the proper interpretation?

While such questions about the interpretation of Scripture were as old as the Church itself, the Church was extremely sensitive to such issues during Galileo's day given the aftermath of the Protestant Reformation. Exactly who had the authority to interpret Scripture was one of the major theological issues that had come between the Church and the Protestant reformers. In fact, some historians have argued that had Galileo made his arguments a century early, he would have met with much less resistance from the Church. (However, given his bombastic writing style and his penchant for making enemies, he may still have been able to find disfavor even then.)

The second concern had to do with the importance and influence of Aristotelian philosophical thought within the Church at the time. For various reasons, Aristotle held that the Earth was at the center of the universe and that the heavenly bodies moved in perfect circles above the Earth on concentric crystalline spheres. The circular motion on the spheres was considered the perfect motion, one that was eternal and incorruptible. The perfection of the heavens seemed to mesh nicely with the Catholic notion of a perfect and eternal Creator, and so the Aristotelian (or Ptolemaic) geocentric system had become intimately tied to Church philosophy and theology. If the Earth was moving around the sun, as Galileo claimed, this would require some significant philosophical and theological reworking.

Given these concerns, coupled with a lack of the observation of parallax, many in the Church did not embrace the heliocentric theory. Despite this, the Church did give Galileo the freedom to present his theory as useful for astronomical prediction provided that he did not claim that it was an actual description of physical reality. Cardinal Robert Bellarmine and Pope Urban VIII, the pope who ultimately had Galileo put on trial and sentenced, both shared this view. In fact, Galileo's famous *Dialogue Concerning the*

Two Chief World Systems (1632), the book that ultimately sealed his fate, was written and published at the request of Pope Urban, who had asked Galileo to write a book giving the arguments for and against heliocentrism. In the book, though, Galileo openly advocated for the heliocentric theory and at the same time seemed to insult Pope Urban by putting some of Urban's ideas on the matter into the mouth of Simplicio, the slow-witted defender of the geocentric system. This move seemed to seal his fate, and shortly after publication, Galileo was called to Rome to stand trial. As a result of the trial, his book was banned, and he was forced to recant his position, one that was deemed to be opposed to Scripture. Galileo spent the remainder of his life under house arrest in a villa near Florence.

While Galileo's fate was certainly lamentable, the Church did eventually reconcile itself with Galileo and his work. The process began in 1741 when the Church allowed the publication of a slightly edited version of the *Dialogue*, and it continued through 1835 when the Church removed the unedited book from the Index of Prohibited Books altogether. Interestingly, the removal of the book from the list occurred two years before parallax was conclusively observed by astronomers.

The Galileo case was an unfortunate incident in the Church's history. It was not, however, a simple matter of the Church oppressing science it did not like. Rather, it was the result of the Church acting with skepticism toward a new and contested theory[5] that clashed with the traditional interpretation of certain scriptural passages. This initial skepticism by some in the Church led to outright condemnation when Galileo's brazen personality turned many of his supporters against him.

It is important to note that the initial skepticism and condemnation of Galileo's ideas was followed by a period of conditional or

5. In fact, the theory was contested by many of the leading scientists of the day for scientific reasons. For a good overview of the scientific debates of the time, see Christopher Graney, "An Alternative History of Modern Science," *Church Life Journal*, January 14, 2020, https://churchlifejournal.nd.edu/articles/an-alternate-history-of-modern-science/.

at least tacit acceptance. This eventually led to a complete acceptance by the Church as marked by the removal of the *Dialogue* from the Index of Prohibited Books. This acceptance occurred well after the theory had become established within the scientific community.

Interestingly, the Church's response to the theory of human evolution demonstrates a similar trajectory, albeit one lacking a sensational trial of a prominent scientist. At the onset, many in the Church responded skeptically to the notion of human evolution, which, like Galileo's theory, was still being debated in scientific circles. Over time, though, as evidence accumulated, the Church moved toward an openness to dialogue regarding the evolution of the human body and its implications for Catholic theological understanding.

When examining the Church's response to evolution, it is important to recognize that the major issue has always been the evolutionary origin of humans. While there are exceptions, there has never been major opposition within the Church regarding the idea that plant and animal species had evolved or that they shared a common ancestry. The Catholic philosopher Kenneth Kemp, who has written extensively on the history of the Church and evolution, points out that "Catholic encyclopedias by the [late nineteenth century] were already stating that the evolutionary origin of plants and animals was not a matter of theological concern."[6]

Rather, what fueled the vast majority of the controversy was the evolutionary connection of man to other animals and the nature of the human soul. From the very beginning, the Church rejected the idea that the soul evolved or was some type of emergent phenomenon of matter. In addition, many in the

6. Kenneth Kemp, "A Brief History of Catholic Evolutionism," Society of Catholic Scientists, May 1, 2021, https://catholicscientists.org/articles/a-brief-history-of-catholic-evolutionism/.

Church were initially skeptical of the idea that even the human body could have evolved from other primates.

While many interpret the Church's initial skepticism and caution toward human evolution as anti-scientific, it is important to put this skepticism into context. It is useful to remember that when Darwin first published *On the Origin of Species* in 1859, his evolutionary ideas were greeted with skepticism even from members of the scientific community who, though sympathetic to the general notion of evolution, thought his theory lacked sufficient explanatory power. Given the robust scientific debate that occurred during the latter half of the nineteenth century regarding various aspects of evolutionary theory, a modicum of caution regarding human evolution was likely the prudent path. A related factor that encouraged caution and skepticism was the decidedly anti-religious bent of many prominent evolutionists of the time, such as Ernst Haeckel and Thomas Huxley.

Making matters worse, Darwin's theory quickly became intertwined with certain modernist philosophical ideas that rose to prominence in the late nineteenth century (much like Galileo's case was affected by the Church's embrace of Aristotelianism and its historical response to the Reformation), making it difficult to engage the science directly without sowing confusion. Finally, the evolutionary origin of the human body posed challenges for the interpretation of certain scriptural passages, and so any hope of integrating human evolution with Catholic theology required a careful scholarly evaluation of passages in Scripture, particularly the first chapters of Genesis. In retrospect, the Church's decision to proceed with caution seems eminently reasonable given both the cultural milieu of the time and the gravity of the matter at stake.

CHURCH TEACHING AND CHURCH DOCUMENTS

In discussing the Church's response to evolution and evaluating what the Church has stated on the matter, it is worth highlighting the different levels of magisterial weight that are attached to various types of Vatican documents or papal addresses. For example, a speech during a general audience that discusses evolution clearly does not carry the same weight as an encyclical. Likewise, an encyclical does not carry the same weight as an apostolic constitution or motu proprio, as the latter tend to be more legalistic and deal with serious doctrinal or administrative matters.

In the case of evolution, it turns out that most of the Vatican documents related to the subject involve either papal exhortations and addresses or documents put forth by papal commissions. These types of documents and addresses fall below encyclicals in terms of their relative weight, as they tend not to discuss matters that are binding on the faithful. For example, papal exhortations or addresses tend to be pastoral in nature and deal with specific issues often addressed to specific groups, while encyclicals tend to be more developed as they shed light on specific Church teachings at specific points in history and are part of the pope's ordinary teaching authority. However, not everything that is stated in an encyclical has equal weight, and its teachings, unless explicitly stated, are not definitive. As a result, the relative weight of any statement within a Church document depends not only on the type of document but also on the context of the statement within the document.

As the Congregation for the Doctrine of the Faith, under the leadership of Cardinal Joseph Ratzinger, articulated, determining the binding nature or importance of a statement within a Church document requires critical analysis. Any specific assertion within a document "require[s] degrees of adherence differentiated according to the mind and the will [the magisterium has] manifested;

this is shown especially by the nature of the documents, by the frequent repetition of the same doctrine, or by the tenor of the verbal expression."[7]

As a result, one must carefully evaluate Church documents to properly understand how binding the various claims/statements in the documents are on the faithful. *Humani Generis*, the only Church encyclical that specifically addresses evolution, provides a good illustration of the need for such analysis. Some of the points in *Humani Generis* involve dogmatic positions held by the Church, such as the claim that man is more than a material being (therefore evolution cannot explain the totality of man). Not only did Pope Pius XII clearly state in *Humani Generis* that "the Catholic faith obliges us to hold that [human] souls are immediately created by God,"[8] but this position has also been reiterated in various pontifical documents and papal addresses on evolution by Popes John Paul II and Benedict XVI.

On the other hand, certain points in the document were not definitive but were proposed for serious consideration given the pastoral needs of the Church at the time. In *Humani Generis*, the suggestion that the body of man evolved from pre-existing matter as well as the concern about theories of polygenism, the idea that mankind emerged as a population rather than a single couple, would fall into this category. In fact, Pope Pius XII's concern regarding polygenism, that "it is in no way apparent how such an opinion can be reconciled with that which the sources of revealed truth and the documents of the Teaching Authority of the Church propose with regard to original sin,"[9] was stated in a qualified manner and has not been repeated in any subsequent Church document. This does not mean that one can ignore the

7. Congregation for the Doctrine of the Faith, "Doctrinal Commentary on the Concluding Formula of the *Professio Fidei*" 11, *L'Osservatore Romano*, July 15, 1998, https://www.ewtn.com/catholicism/library/doctrinal-commentary-on-concluding-formula-of-professio-fidei-2038.

8. Pius XII, *Humani Generis* 36, encyclical letter, August 12, 1950, vatican.va.

9. *Humani Generis* 37.

concerns Pope Pius XII raised about polygenism, but it suggests the document does not provide definitive Church teaching on the topic.

In fact, most of the papal and Church documents related to evolution are not intended to provide definitive Catholic teaching on the most controversial aspects of the evolution/creation question, like the origin of man, the nature of the fall and original sin, and the proper interpretation of Genesis. However, the mere fact that these documents do not provide definitive Church teaching on these matters does not mean that they can simply be ignored. Rather, they can still provide useful guidance for approaching these contentious issues. In particular, the tenor of recent Church and papal documents on the matter have displayed a greater receptivity to the science of evolution than earlier documents. Furthermore, these documents have suggested that evolution and the Church's understanding of the origin of man are not inherently incompatible.[10]

THE BEGINNINGS OF A CONTROVERSY

When Darwin published his *Origin of Species* in 1859, it caused quite a stir among religious believers. Although the story is likely apocryphal, the response attributed to the Bishop of Worcester's wife when she first heard of Darwin's theory of evolution is telling: "Descended from the apes! My dear, let us hope that it is not true, but if it is, let us pray that it will not become generally known." Even several Catholics were quick to denounce as heretical the idea that man's body had evolved from other primates. One of the most influential Catholic theologians of the time, Matthias Joseph Scheeben, specifically stated that it was heretical to postulate that man's body "is descended from monkeys' as a consequence of a

10. It's important to remember that this is not an indication that the Church "endorses" the scientific theory of evolution. The Church does not endorse scientific theories. However, this situation does indicate that there is not something inherently scandalous in evolutionary theory.

progressive change registered in forms."[11] Despite the concerns of Scheeben and many others, like the Jesuit theologian Camillo Mazzella, the Vatican never publicly commented on the evolutionary origin of man throughout the nineteenth century, despite the significant opposition toward it by many Church officials.[12]

The earliest reference to human evolution in Church documents is to be found in a decree from the Provincial Council of German Bishops in Cologne in 1860. The document stated that the "spontaneous transformation" of a primate into a human was deemed contrary to the faith. However, many have argued that such a statement does not dismiss human evolution out of hand; rather, it only condemns a spontaneous transformation in which God plays no active part. As pointed out by Mariano Artigas in his book *Negotiating Darwin: The Vatican Confronts Evolution, 1877–1902*, even with this condemnation, there would not necessarily be a problem "in accepting evolution so long as one recognized simultaneously the necessity of divine participation for the process to take place, in such a way that the secondary causes [evolutionary mechanisms] might join with continuous divine action [primary causation] in giving being and activity to all organisms."[13] Regardless of this distinction, the position of the Provincial Council was never endorsed by the Vatican, so it did not reflect official Church teaching on the subject.

For his part, Pope Pius IX, the pontiff at the time, was concerned about the spread of Darwin's evolutionary ideas throughout Italy during the 1860s. His concern largely stemmed from the fact that evolution was being used to support philosophical materialism, a position that implied man was nothing more than a material being and that is clearly at odds with the faith. Despite this legitimate concern, Pope Pius IX never placed

11. Matthias Joseph Scheeben, *Handbuch der katholischen Dogmatik*, book 3, Schopfungslehre, 160–161, reprinted in Mariano Artigas et al., *Negotiating Darwin: The Vatican Confronts Evolution, 1877–1902* (Baltimore, MD: Johns Hopkins University Press, 2006), 19.

12. Artigas et al., 19–23.

13. Artigas et al., 23.

Darwin's *Origin* onto the Vatican's Index of Prohibited Books, nor did he make any official pronouncement regarding the theory.

As in the Galileo case, Pope Pius IX's concerns regarding evolution were tied up with the major historical issues the Church was facing at the time—namely, the influence of nineteenth-century thinkers who argued for the elevation of reason above faith, or worse yet, the complete separation of faith and reason. In this climate, many scholars were arguing that both philosophy and modern science should pursue their studies with no reference to the truths of the faith, a position that led to the spread of many ideas that were contrary to the teachings of the Church. Materialistic evolutionary ideas were just one aspect of this cultural juggernaut, a movement that spawned atheistic philosophies, the closure of convents and monasteries, increasing levels of anticlericalism, and "scholarly" scriptural interpretations that were clearly contrary to the Catholic faith.

In 1869, Pope Pius IX convened the First Vatican Council in part to address the crisis. While there were much larger issues at stake, the work of the council had relevance to the Church's engagement with evolution. Of particular interest regarding evolution is the Vatican I dogmatic constitution *Dei Filius*. This document articulated the basic Catholic principle—a principle that can be traced back to the Scholastic philosophers—that would underlie the Church's effort to address the evolution/creation issue in the years ahead: "But, although faith is above reason, nevertheless, between faith and reason no true dissension can ever exist, since the same God, who reveals mysteries and infuses faith, has bestowed on the human soul the light of reason; moreover, God cannot deny Himself, nor ever contradict truth with truth."[14]

The notion that truth cannot contradict truth implies that any apparent contradictions between the truths of the faith and those truths acquired through the use of human reason are the

14. Vatican Council I, *Dei Filius* 4, dogmatic constitution, April 24, 1870, inters.org.

result of either the teachings of the faith being improperly understood or human reason being improperly applied. If one truly understood the faith and used right reason, any apparent contradictions between faith and science could eventually be resolved, at least in principle. On the contrary, if one merely pursued a scientific study of the origins of man without any reference to the truth of man as revealed by faith, or a scholarly critique of the Genesis creation story without any reference to Church teachings on Scripture and creation, one would be in danger of reaching errant conclusions.

While this approach set the stage for the Church's engagement with modern science, particularly in the twentieth century, the council did not directly address the major issue surrounding human evolution—namely, how one was to properly interpret Genesis in relation to the possible evolutionary origin of man. This issue was touched upon indirectly by Pope Leo XIII in his 1893 encyclical *Providentissimus Deus* and later by the Pontifical Biblical Commission in the early 1900s. In his encyclical, Pope Leo XIII made clear that the truth of Scripture could not contradict the truth of science and any apparent discrepancy should be reconciled by careful study. While he did not advocate a nonliteral reading of Genesis, he did note that one could dispense with a literal-historical reading of Scripture and even "push inquiry and exposition beyond what the Fathers have done; provided [one] carefully observes the rule so wisely laid down by St. Augustine—not to depart from the literal and obvious sense, except only where reason makes it untenable or necessity requires."[15]

The Pontifical Biblical Commission, established by Pope Leo XIII in 1902 to ensure the proper interpretation of Scripture, made a similar assertion. The commission brought attention to the fact that the Church Fathers had applied nonhistorical interpretations to various parts of Genesis. In addition, when it came to the Genesis creation accounts, the Church Fathers put

15. Leo XIII, *Providentissimus Deus* 15, encyclical letter, November 18, 1893, vatican.va.

forth a variety of interpretations.[16] Some Fathers believed that everything had been created simultaneously rather than over six days. Others, focusing on the words in the Genesis text "let the earth bring forth," believed that life had come forth in an ordered fashion from the earth rather than being specifically formed out of nothing by God. Still others thought that the heavenly bodies were instilled with life-generating powers to bring forth living forms on the earth.

Given the wording of *Providentissimus Deus* and the variety of interpretations by the Fathers, there was no definitive Church teaching insisting that the Genesis creation story be taken as a historical description of events. In principle, the Church was open to interpretations of Genesis that were more compatible with evolutionary theories, although the Pontifical Biblical Commission in 1909 stated that the faithful must not deny the special creation of man.[17]

EARLY CATHOLIC RESPONSES TO EVOLUTION

Just as in the Galileo case, the Church's early response to evolutionary ideas has to be placed in a historical context to be properly understood. The fact that Galileo was advocating for his theory during the Counter-Reformation period likely affected the Church's response to his work. Similarly, the Church's initial lukewarm (and sometimes hostile) response to evolutionary theory must be seen in the context of the events of the late nineteenth century. At this time, when Darwin's theory was initially being debated among the scientific community, biblical scholars had begun to apply what is called the historical-critical method to understanding the Bible. This method focused on the historical nature of the text, asking such questions as who wrote it,

16. The interpretations of some of the Church Fathers are discussed in more detail in chapters 3 and 8.

17. Pontifical Biblical Commission, "Concerning the Historical Nature of the First Three Chapters of Genesis," June 30, 1909.

what culture did they inhabit, what genre did they use, etc., to better understand the text. Such a method had the ability to deepen the Church's understanding of Scripture, but it also raised the possibility that the Bible could be reduced to just another human-authored ancient text. Such a view could easily overshadow the reality that the Bible was the inspired Word of God. The Church was concerned that if one *only* focused on the human dimension of Scripture, it could lead to many speculative theories that might undermine the central tenets of Christian doctrine, including the Incarnation, creation, and redemption. For example, some scholars were arguing that the meaning of biblical texts and the sacred deposit of faith could evolve over time and change meaning entirely based upon the current audience. (This stands in contrast to the notion of the development of doctrine that St. John Henry Newman discussed at the time, in which a deeper and clearer understanding of a specific doctrine could indeed develop over time.) If such a position were correct, it meant that essential aspects of the faith, like the belief in the divinity of Christ, were subject to debate.

The Church had good reason to be concerned. On the one hand, it seemed that modern scholarship could help in properly understanding certain scriptural passages such as the Genesis text. But biased and agenda-driven scholarship—reason gone amok—had the possibility of eroding the faith by encouraging the uncritical reinterpreting of all Scripture in light of secular ideologies that were prevalent at the time.

In addition, scholars had begun to extend evolutionary ideas to disciplines ranging from economics to sociology. In a rather short period of time, biological theories of evolution as applied to man were used to justify everything from eugenics to atheism. Given these associations, and the fact that materialistic theories of evolution conflicted with the belief in the existence of man's immaterial soul, many Catholics were strongly opposed to evolution and produced books and pamphlets seeking to undermine it. In particular, the Jesuit publication *Civilta Cattolica* frequently published articles attacking evolution. These attacks were often scientific in nature, picking up ammunition from the criticism Darwin's theory was facing from within the scientific community in the late 1800s and early 1900s.[18]

Such suspicion and hostility particularly toward human evolution among Catholics in the late nineteenth century was by no means universal. For example, St. John Henry Newman seemed sympathetic to the idea that man's body had evolved from other primates. Newman stated that the fact that primates had such similarities with humans seemed to suggest some type of evolutionary relationship. In addition, other prominent Catholics such as the scientist St. George Jackson Mivart[19] and the American priest John Zahm wrote books about the compatibility of evolution and the Catholic faith. The Vatican's response to these men varied considerably, demonstrating that there was still no official Vatican position on many of the issues at stake.

Mivart published *On the Genesis of Species* in 1871, and in it he stressed the distinction between primary creation, the creation of something out of nothing by God, and derived or secondary creation, the formation of something from the creative powers that God has instilled in his creation. In this manner, he attempted to reconcile evolution—secondary creation—with God's primary act of creating the universe out of nothing. Recognizing this distinction, he made clear that one could be an evolutionist and a Catholic so long as one admitted that the soul was immaterial and created directly by God. His work was well received by Pope Pius IX, who bestowed upon him an honorary doctorate of philosophy five years after the publication of *On the Genesis*

18. While Darwin's theory was vigorously debated in the scientific community at the time, it is important to recognize that the debate was not regarding evolution in general nor the idea that species were related by common descent. In fact, these views had become widely accepted within the scientific community during the latter portion of the nineteenth century. Rather, the debate centered on whether natural selection could suffice as an explanation for evolutionary change.

19. Despite his name, Mivart is not a canonized saint of the Catholic Church.

of Species. In his later years, Mivart did run into trouble with the Vatican regarding his views on hell, but the Vatican never censured his views regarding evolution, nor did it ever place *On the Genesis of Species* on the Vatican's Index of Prohibited Books.

Other Catholics who wrote about evolution found that their writing was not as well received. For example, Fr. John Zahm's 1896 book *Evolution and Dogma*, which discussed human evolution along much the same lines as Mivart, met with considerably more resistance. Despite the similar line of argumentation, Fr. Zahm was forced to have his publisher withdraw his book from the market to placate the concerns of the Vatican. (Despite being forced to remove the book from circulation, there was no public condemnation of *Evolution and Dogma* by the Vatican and the book never appeared on the Index of Prohibited Books.) Why then did his book meet a completely different fate than Mivart's? Scholars have argued that Fr. Zahm's troubles with the Vatican had more to do with his role in the Americanist movement in the Church, a role that created enemies for him in Rome, than with the book's content. However, the degree to which his involvement in the movement to modernize the Church in America affected the reaction to his book in Rome is unclear. What is clear, though, is that there was no consistent pattern by which the Vatican at the time dealt with works of speculative theology regarding evolution. However, a number of other Catholic books on the subject of evolution and creation did meet with a similar fate as Fr. Zahm's *Evolution and Dogma*, quietly being removed from circulation yet never ending up on the Index.[20] As a result, the spread of evolutionary ideas among Catholics at the turn of the century was slowed without the Vatican issuing any public statements denouncing evolution.

In some respects, it seems surprising that the Church did not react in a harsher manner to evolution given the threat

20. For a good overview of the fate of such books during this time period, see Artigas et al., *Negotiating Darwin*.

materialistic versions of evolution posed to Catholic theology. The Church certainly came down hard against many other lines of modern thought in the late 1800s. Why then did it not come down harder on evolution, particularly theories regarding the evolution of man?

To some extent, one could argue that the Church's cautious and measured response to this issue was the only reasonable option given the circumstances. On the one hand, the Church had deep reservations and doubts about a relatively new scientific theory that was coupled with dubious modern philosophical ideas. On the other, it did not want to risk furthering the impression that it was opposed to scientific discovery, an impression that was gathering steam at the time due to popular but since debunked polemic tracts such as John William Draper's *History of the Conflict Between Religion and Science*, a book that framed traditional Christian faith as *the* major obstacle to the progress of science. The Vatican's public silence on the matter, coupled with its efforts to persuade quietly a number of Catholic authors to withdraw their books on the subject, seemed to maintain a tenable middle ground in the face of a new and uncertain theory.

THE VATICAN DIRECTLY ENGAGES EVOLUTION

The first half of the twentieth century saw most scientists accept Darwin's theory as the best explanation for the origin of species, including man. While there remained a significant minority of scientists who were skeptical of Darwin's theory of natural selection, most of these skeptics were still evolutionists. They simply differed by the fact that they argued that other natural processes were the primary drivers of evolution. In addition, evolutionary theory had become a standard part of the teaching at the university level, making it all the more imperative that the Church address the subject directly.

As evolutionary theory rose to prominence in the scientific

community, books that attempted to reconcile evolution and the Catholic faith found less opposition from the Church. Henri de Dorlodot's 1921 work *Darwinism and Catholic Thought* and Ernest Messenger's 1931 work *Evolution and Theology: The Problem of Man's Origin* enjoyed a relatively favorable reception within the Church. Both authors were Catholic clergy, and both looked with favor on the theory that the human body could have evolved from lower life forms. They did, however, maintain that this had yet to be demonstrated conclusively by scientists.

Other theologians, however, ventured more speculative ideas regarding the theology of evolution during this period. The most famous was the French Jesuit Pierre Teilhard de Chardin, who argued that sin was the natural outcome of an evolutionary process in which creation was striving toward perfection, rather than the consequence of man's original fall from grace. According to Teilhard de Chardin, the doctrine of original sin was not compatible with evolution and therefore needed to be revised. He was eventually censured by the Church for these views.

The appearance of these more speculative ideas among some Catholic theologians, coupled with the widespread acceptance of evolution within the scientific community, seemed to give the Vatican the impetus to discuss the issue of evolution head on. Pope Pius XII began this process in a 1941 address to the Pontifical Academy of Sciences, a body he was instrumental in establishing during the previous papacy of Pius XI.

In his address, Pope Pius XII discussed the origin of man from a theological perspective and alluded to the science of evolution. While he seemed to acknowledge some tension between these concepts, he did not attempt to resolve it. In discussing the origin of man, he stated that "man [was] fashioned out of dust from the soil and God breathed into his nostrils a breath of life and thus man became a living being."[21] This is then followed by a

21. Pius XII, "God the Only Commander and Legislator of the Universe," Address to the Pontifical Academy of Sciences, November 30, 1941, https://www.ewtn.com/catholicism/

line that has suggested to some that the pope was placing humankind entirely outside the scope of an evolutionary process: "Only from man could there come another man who would then call him father and ancestor."[22]

However, Pope Pius XII also indicated that the science regarding man's physical origin is uncertain and that we must leave it to future researchers to uncover the truth: "The multiple research, be it palaeontology or of biology and morphology, on the problems concerning the origins of man have not, as yet, ascertained anything with great clarity and certainty. We must leave it to the future to answer the question, if indeed science will one day be able, enlightened and guided by revelation, to give certain and definitive results concerning a topic of such importance."[23] It seems clear from this line in the text that the pope's address to the Academy was not meant to settle the matter among Catholics with respect to the evolution of the human body. Less than ten years later, though, Pope Pius XII did provide some clarity regarding human evolution and its relation to the faith in the form of his 1950 encyclical *Humani Generis*, the first official papal document to address specifically the scientific theory of evolution.

During his pontificate, Pope Pius XII stressed continually that scientific truth could not be in conflict with the truths of the faith, reiterating the long-standing position of the Church. Pope Pius XII believed that the Church had nothing to fear from modern science if scientific discovery was pursued free from overt prejudice and bias. Despite this openness toward science in general, the encyclical was far from a ringing endorsement of evolution. It did state, though, that the Catholic Church was open to the possibility, albeit conditionally, that the evolution of the human body had in fact occurred. The encyclical states

library/to-plenary-session-of-the-pontifical-academy-of-sciences-8957.

22. "God the Only Commander and Legislator of the Universe."

23. "God the Only Commander and Legislator of the Universe."

the following: "The Teaching Authority of the Church does not forbid that, in conformity with the present state of human sciences and sacred theology, research and discussions, on the part of men experienced in both fields, take place with regard to the doctrine of evolution, in as far as it inquires into the origin of the human body as coming from pre-existent and living matter—for the Catholic faith obliges us to hold that souls are immediately created by God."[24]

Immediately after stating this, Pope Pius XII voiced concerns regarding the evolutionary idea of polygenism, the notion that humans had not originated from a single set of parents. Polygenism was, and is, the leading scientific view of human origins, a view that posits modern humans evolved from a population or group of pre-human primates rather than just originating from a single couple. Pope Pius XII's concern with polygenism stems from the fact that he did not see how this could be made consistent with the Catholic understanding of original sin, a sin that is passed on to the entire human race from our first parents.[25]

Even though he was concerned with the notion of polygenism, the mere admittance of evolution as a matter worth discussing opened up more room for Catholic theological work. By the time the Second Vatican Council convened in 1962, significant work had been done by theologians in this area. Such theology had an indirect effect on Vatican II even though biological evolution was never addressed in its documents. Despite this omission, the documents of Vatican II have much to offer regarding the integration of evolution and the Catholic faith. First, by reiterating what was articulated at Vatican I, the documents make clear that there can be no real conflict between science and faith. Second, while biological evolution was never specifically mentioned, there was acknowledgment of the fact that "the human race has passed

24. *Humani Generis* 36.

25. Chapter 9 is dedicated to a discussion of original sin in light of an understanding of human evolution. More on this topic will be discussed there.

from a rather static concept of reality to a more dynamic, evolutionary one."[26] Third, the documents encouraged careful biblical scholarship aimed at understanding the intention of the sacred writers, implying that not all passages were to be seen as historical narratives. In this biblical scholarship, "attention should be given, among other things, to 'literary forms.' For truth is set forth and expressed differently in texts which are variously historical, prophetic, poetic, or of other forms of discourse."[27] Finally, the council's documents displayed a congenial attitude toward natural science in general, both speaking favorably regarding its progress in understanding the natural world and acknowledging its rightful autonomy. At the same time, the documents explicitly warned of the dangers of "transgressing the limits of the positive sciences"[28] and reaching the erroneous materialistic conclusion that science can explain all of reality.

EVOLUTION AND THE INTERPRETATION OF GENESIS

As the Vatican II documents suggest, the question of how to properly integrate creation and evolution is closely related to the question of how to properly interpret the first few chapters of Genesis. Much more will be discussed in relation to this question in the next chapter, but it is worth pointing out how the Church's guidance on this issue has developed since the time of Darwin.

One of the most explicit references (post-Darwin) regarding how to interpret the Genesis creation accounts can be found in the 1909 Pontifical Biblical Commission (PBC) document "Concerning the Historical Character of the First Three Chapters of Genesis." In this text, the PBC states that one must not doubt the historical sense of "the creation of all things by God in the

26. Vatican Council II, *Gaudium et Spes* 5, in *The Word on Fire Vatican II Collection: The Constitutions*, ed. Matthew Levering (Park Ridge, IL: Word on Fire, 2021), 220.

27. *Dei Verbum* 12, in *Vatican II Collection: The Constitutions*, 28.

28. *Gaudium et Spes* 19, in *Vatican II Collection: The Constitutions*, 234.

beginning of time; the special creation of man; the formation of the first woman from the first man; the unity of the human race; the original felicity of our first parents in the state of justice, integrity, and immortality; the command given by God to man to test his obedience; the transgression of the divine command at the instigation of the devil under the form of a serpent; the degradation of our first parents from that primeval state of innocence; and the promise of a future Redeemer."[29]

To clarify the magisterial weight of the PBC document, Pope Pius X in his motu proprio *Praestantia Scripturae* in 1907 stated that Catholics were "bound by the duty of conscience to submit to the decisions of the Biblical Pontifical Commission, both those which have thus far been published and those which will hereafter be proclaimed, just as to the decrees of the Sacred Congregations which pertain to *doctrine* and have been approved by the Pontiff."[30] While the belief that God created all things is a matter of doctrine, the belief that woman was formed from man or that our first parents were tricked by the devil under the form of a serpent are not doctrinal teachings of the Church, so not all aspects of the PBC document would hold equal weight. In addition, what exactly was meant by a historical understanding was not clarified in the initial document.

It wasn't until four decades later, after Pope Pius XII asked the PBC to reexamine the question given the state of modern science and biblical scholarship, that some clarity was given. In 1948, the commission produced, and Pope Pius XII approved, a much more nuanced statement regarding the proper historical interpretation of the text. The PBC pointed out that determining the historical meaning of Genesis, the key issue in the 1909

29. "Concerning the Historical Character of the First Three Chapters of Genesis," June 30, 1909, in *A Catholic Commentary on Holy Scripture*, ed. Bernard Orchard and Edmund F. Sutcliffe (Toronto: Thomas Nelson, 1953), 69.

30. Pius X, *Praestantia Scripturae*, motu proprio, November 18, 1907, in *The Sources of Catholic Dogma*, ed. Henry Denzinger and Karl Rahner, trans. R.J. Deferrari (St. Louis, MO: Herder, 1954), 543 (emphasis added).

document, was a difficult exercise because our modern categories of literary types do not readily apply to the first couple of chapters of Genesis. It stated the following: "These literary forms do not correspond to any of our classical categories and cannot be judged in the light of the Greco-Latin or modern literary types. It is therefore impossible to deny or to affirm their historicity as a whole without unduly applying to them norms of a literary type under which they cannot be classed. If it is agreed not to see in these chapters history in the classical and modern sense, it must be admitted also that known scientific facts do not allow a positive solution of all the problems which they present."[31]

Given the situation—known scientific facts that appear to conflict with a historical reading (in the modern sense) as well as difficulties in properly understanding these ancient literary forms—the proper interpretation of the text was not readily apparent. In response to this conundrum, the PBC, mirroring what Pope Pius XII would state in *Humani Generis*, stressed the need for continued work and collaboration among those formed in various scientific and theological disciplines:

> The first duty in this matter incumbent on scientific exegesis consists in the careful study of all the problems literary, scientific, historical, cultural, and religious connected with these chapters; in the next place is required a close examination of the literary methods of the ancient oriental peoples, their psychology, their manner of expressing themselves and even their notion of historical truth; the requisite, in a word, is to assemble without preformed judgements all the material of the palaeontological and historical, epigraphical and literary sciences. It is only in this way that there is hope of attaining a

31. Pontifical Biblical Commission, "Regarding the Sources of the Pentateuch and the Historical Value of Genesis 1–11," January 16, 1948, vatican.va.

clearer view of the true nature of certain narratives in the first chapters of Genesis.[32]

This cautious attitude, which was also reiterated in *Humani Generis*[33] and other recent documents, seems to be part of the natural response to scholarly developments in the twentieth century. In the case of evolution, as more evidence has been presented suggesting that humans have evolved from other primates, the Church has encouraged scholars of good faith to work together to properly glean insights from the Genesis text while at the same time cautioning them against rash and overly speculative interpretations. In this manner, the Church has provided guidance regarding how to proceed in studying Genesis considering evolutionary science while refraining from providing definitive exegetical judgments. Hence, the Church encourages fruitful discussions regarding the "true nature of certain narratives in the first chapters of Genesis."

CREATION AND EVOLUTION

While the Church has not provided definitive guidance regarding how Catholics must approach the evolution/creation issue, there has been a concerted effort to integrate these two concepts since the publication of *Humani Generis*. This openness is evident in probably the most famous address any pope has ever made regarding evolution, John Paul II's 1996 address to the Pontifical Academy of Sciences. In the address, he spoke of the progress of scientific studies regarding evolution and the implications this had for the Church: "Today, nearly half a century after the publication of [*Humani Generis*], new knowledge leads to the recognition of

32. "Regarding the Sources of the Pentateuch."

33. Pope Pius reiterated the point that while the first chapters of Genesis present history in a manner "not conforming to the historical method used by the best Greek and Latin writers or by competent authors of our time, [they] do nevertheless pertain to history in a true sense, which however must be further studied and determined by exegetes" (38).

the theory of evolution as more than a hypothesis. It is indeed remarkable that this theory has been progressively accepted by researchers following a series of discoveries in various fields of knowledge. The convergence, neither sought nor provoked, of the results of work that was conducted independently is in itself a significant argument in favor of this theory."[34]

While he never explicitly endorsed any specific scientific theory of evolution, something that would be quite strange for a pope to do, he did explicitly reject evolutionary ideas that, "in accordance with the philosophies inspiring them," denied the existence of a human soul that is created immediately by God. In this respect, his address was consistent with what Pope Pius XII articulated in *Humani Generis*.

What made his speech different than *Humani Generis* was the tone in which John Paul II referred to evolution. Rather than the cautious language seen in *Humani Generis*, John Paul II displayed a sense of appreciation at the remarkable amounts of research from different fields of science that supported the broad concept of evolution. This was a significant change from the more guarded manner by which the Church had viewed evolution throughout most of the twentieth century, and it did not go unnoticed.[35] Press headlines from the time read "Pope Endorses Evolution." While this statement was not quite true, as popes are not in the business of endorsing scientific theories, he did make clear that the Church was ready to engage with the *scientific truths* that researchers had uncovered regarding evolution.

What then did John Paul II make of Genesis? If the evolutionary origins of man seemed possible via a reasoned view of the science, how was one to interpret the Genesis account of creation? In his earlier writings, John Paul II was quite clear that Genesis was not meant to be a historical description of the

34. John Paul II, "On Evolution" 4.

35. This did not mark any type of change in Church teaching on the matter, merely a change in attitude. Despite the openness both John Paul II and Benedict XVI had to evolution, neither established any new definitive magisterial teachings on the matter.

creation process: "The Bible itself speaks to us of the origin of the universe and its makeup, not in order to provide us with a scientific treatise. . . . Sacred Scripture wishes simply to declare that the world was created by God, and in order to reach this truth it expresses itself in the terms of the cosmology in use at the time of the writer. . . . Any other teaching about the origin and make-up of the universe is alien to the intentions of the Bible."[36]

According to John Paul II, the text is not in conflict with science because the Genesis text is attempting to articulate truths about what the created world is—namely, the product of God's handiwork—and not deliver a scientific treatise on the process by which God created it. The same can be said regarding the origin of man. The Bible describes what man is—a creature made in the image and likeness of God—and does not describe the scientific or biological process that contributed to the origin of man. This interpretation of the Genesis creation account will be developed more in chapter 4, and it is an interpretation of Genesis that John Paul II shared with his successor, Pope Benedict XVI.[37]

Pope Benedict XVI unarguably took a greater interest in evolution and its relationship to the Genesis creation account than any other pontiff. In 1986, while still a cardinal, he published a book entitled *"In the Beginning . . .": A Catholic Understanding of the Story of Creation and the Fall* that explored the relationship between the Genesis text and creation. He oversaw the writing of Church documents on the subject and even organized a colloquium with his former students to discuss the issue, the proceedings of which were published as a book entitled *Creation and Evolution*. He even referred to evolution in the sermon at the inauguration Mass of his pontificate, stating, "We are not some casual and meaningless product of evolution. Each of us is the

36. John Paul II, "Cosmology and Fundamental Physics," Address to the Pontifical Academy of Sciences, October 3, 1981, https://www.ewtn.com/catholicism/library/cosmology-and-fundamental-physics-8135.

37. This is not to say that it is the only possible Catholic interpretation. It is, however, a valid Catholic interpretation of the text.

result of a thought of God."[38] Clearly, it is a topic about which he spent much time deliberating.

While the quote from his inaugural Mass makes it sound like he is critical of evolution, the reality is that his position on the matter was very similar to John Paul II's. Like his predecessor, Benedict did not see evolution and creation as an either/or proposition, but rather as a both/and scenario. In a key passage from *"In the Beginning,"* he states, "We cannot say: creation or evolution, inasmuch as these things correspond to two different realities. [The Genesis creation account] does not in fact explain how human persons come to be but rather what they are. . . . And, vice versa, the theory of evolution seeks to understand and describe biological developments. But in doing so it cannot explain where the 'project' of human persons comes from, nor their inner origin, nor their particular nature. To that extent we are faced here with two complementary—rather than mutually exclusive—realities."[39]

With both John Paul II and Benedict XVI, there has been a tangible change in the Church's engagement with science on the topic of evolution. The Church no longer seems warily distant and overly guarded on the matter; rather, it seems eager to engage and understand that which is true and valid regarding the science of evolution. Rather than being on the defensive, the Church is now leading the way and encouraging scientists to enter into a dialogue. Such an integration of science and theology on the subject holds the promise not only to purify science from ideology but to enrich and encourage theological study as well. As John Paul II has asked, "Does an evolutionary perspective bring any light to bear upon theological anthropology, the meaning of the human person as the imago Dei, the problem

38. Benedict XVI, "Homily of His Holiness Benedict XVI," April 24, 2005, vatican.va.

39. Joseph Ratzinger, *"In the Beginning . . .": A Catholic Understanding of the Story of Creation and the Fall* (Grand Rapids, MI: Eerdmans, 1995), 50.

of Christology—and even upon the development of doctrine itself?"[40]

This does not mean that the science of evolution will overturn Church doctrines regarding such things as the dependence of all things on God's creative power, the reality of original sin, or the immateriality of the soul. It may, however, help doctrine develop in richness and depth. This can only occur, though, if the science and religion dialogue is performed within the proper framework, as Pope Benedict made clear in his inaugural address. Like previous popes, Benedict has also warned against those who would equate evolution with errant philosophies that reduce man to a purely material being. An acceptance of the scientific reality of evolution must always be qualified, for there are those who equate evolution with atheism and materialism.

This is an important point to remember. While this book approaches the issue through the framework laid out by Pope Benedict XVI—evolution and creation are two complementary realities—it is critical to be clear about what one means by the words creation and evolution. Gaining a clear understanding of the Catholic dogma of creation as well as the different meanings of the term evolution will be the task of the next two chapters.

40. John Paul II, "Letter of His Holiness John Paul II to Reverend George V. Coyne, SJ, Director of the Vatican Observatory," June 1, 1988, vatican.va.

4

A Catholic Understanding of Creation

Nearly all peoples have developed their own creation myth, and the Genesis story is just the one that happened to have been adopted by one particular tribe of Middle Eastern herders. It has no more special status than the belief of a particular West African tribe that the world was created from the excrement of ants.[1]

—Richard Dawkins

Nearly every culture has its own creation story, a response to the question of how the world came to be and why it is arranged in the manner that it is. The Chinese have a story that involves the giant Pan Gu breaking forth from a huge chaos-containing egg to impart order to the universe. The Iroquois have the legend of a woman falling from the sky and being aided by the birds and fish to establish the Earth as an island on the back of the Great Turtle.

In most of these cases, the creation story is synonymous with the beginning. It is the answer to the question of how the world we see around us got its start. At first glance, the beginning chapter of Genesis seems to answer this question for Christians. It depicts God bringing the world into existence in six successive days as he separates light from dark, heavens from earth, and land from water. But the depiction of creation in Genesis uses this poetic language not necessarily to describe the historical events regarding how the world got its start but to emphasize the radical dependence all things have on God for their very existence. As the *Catechism* points out, the overarching message of the Genesis

1. Richard Dawkins, *The Blind Watchmaker* (New York: W.W. Norton, 1986), 316.

creation account is that "the totality of what exists (expressed by the formula 'the heavens and the earth') depends on the One who gives it being."[2] For Catholics, *this* is the key dogma of creation.

The philosopher William Carroll frames the Catholic principle of creation this way: "To create is to give existence, and all things depend upon God for the fact that they are. . . . Any thing left entirely to itself, separated from the cause of its existence, would be absolutely nothing."[3] His point is that everything that exists in the world, from squirrels to humans to quarks, exists only because God, as the ultimate source of existence, causes it to exist. None of these entities have the ability, on their own, to bring themselves into existence: they are said to be contingent. Contingent entities are dependent upon something else for their very existence, not just at the onset of their existence but at every moment of their existence. Framed in this fashion, God's role as Creator is not limited to a moment that comes and goes. Rather, God sustains everything that exists and holds everything in existence at *every* point in time. If the universe had a beginning, as both modern science and Catholic theology claim, God was responsible for sustaining the existence of whatever was there at the beginning of time. However, he is also responsible for sustaining the existence of everything that is here yesterday, today, and tomorrow.

This Catholic dogma of creation provides an answer to the question of why our universe exists at all, a question that has occupied the brightest minds throughout recorded history. The universe on its own does not seem to provide an adequate answer regarding its existence. In fact, the more we learn about the universe we occupy, the more it becomes apparent that there is nothing that dictates 1) that it had to exist or 2) that it had to exist in the state that we find it. It is an entirely contingent

2. *Catechism of the Catholic Church* 290.

3. William E. Carroll, "Aquinas and the Big Bang," *First Things* 97 (November 1999): 18–20.

entity. What does this mean? It means that it could easily have been otherwise. It could have been a universe with no complex molecules or one devoid of stable laws and forces. More importantly, it could have just as easily never existed. In fact, as Leibniz pointed out, that would have been a much simpler option. Why then does this specific universe exist?

One possible answer to this question is to posit that the universe is not contingent, that for some reason it had to exist out of necessity. For example, some physicists have argued that the universe's existence is the necessary outcome of the interactions of gravity in a quantum vacuum. Others have argued that our universe necessarily came into existence from fluctuations in some type of multiverse. While such explanations for the *beginning* of our universe may or may not be true, they entirely miss the point. First, they only attempt to explain how the universe got started, thereby leaving the existence question untouched. (Why does the universe exist at all?) Secondly, these arguments only attempt to explain the origin of one contingent entity, our universe, by appealing to some other contingent entity, the multiverse or gravity or quantum vacuums. In doing so they only push the existence question back one more level. Rather than why does the universe exist, the question becomes, why does a multiverse exist (if it does)? Or why do gravity and quantum vacuums exist? This fundamental question of existence—why do things exist at all—is one that cannot be answered by merely invoking another level of contingent material entities, like using the multiverse to explain the existence of our universe. It is precisely the question of why *any* contingent material entities exist that one is seeking to answer.

To escape from this conundrum, the philosopher Bertrand Russell argued that the existence of the universe is a brute fact. It just is, and we have no need to explain its existence. His reasoning does not provide an answer to the question of why the universe exists but simply ignores the question as irrelevant. One can take this position, but it is extremely intellectually unsatisfying, as it ultimately fails to provide a reason for the existence of our ordered universe. It simply just is, and that's that.

In contrast, the Catholic dogma of creation addresses this question head on and posits an answer: the ultimate source of the universe is Existence itself—namely, God. Everything else in existence derives its existence from God, who is the only necessary source of existence. This is the foundational Catholic dogma of creation. It is important to note that this position is not taken merely on faith. Using human reason, one can make a strong case in support of this position. Such a case, though, is not a scientific case because claims about ultimate existence are not the type of claims that fit within the scope of scientific inquiry. Modern science cannot properly address why a contingent entity such as the universe exists. However, the Catholic philosophical understanding of creation can, and it can do so in a very robust and reasoned manner. It bears witness to a truth that all can access through the right use of human reason: the world has a Creator who sustains all things in existence at all times. This principle holds whether the universe sprang into existence fully formed or emerged in its present form as the result of an evolutionary process that played out over immense time spans. Either way, the universe owes its existence to God.

GOD'S CAUSALITY

The Catholic dogma of creation makes the claim that God is the ultimate cause of all that exists. However, it is important to recognize that God is an entirely different type of cause than the causes we normally observe operating in the world around us. As pointed out earlier in the book, recognizing this distinction is essential if one is to properly integrate evolution and creation. One of the most common concerns that Christians voice regarding the integration of evolution and creation speaks to this distinction. It is often voiced as a version of the following question: How does one reconcile the notion of a providential God who knows us, loves us, and planned for us with an inherently chance-ridden evolutionary process? For these people, the line from Jeremiah, "Before I formed you in the womb I knew you, and before you were born I consecrated you" (Jer. 1:5), seems at odds with an evolutionary process that by chance produced the human form. In short, the question seems to boil down to this: Did God knowingly and lovingly create us, or are we the chance by-product of an evolutionary process? The implication of this question is that the process of evolution somehow stands in competition with God's causal power. The two are seen as combatants in a zero-sum game in which one triumphs at the expense of the other. However, this false dichotomy only arises if one reduces God to just one cause among others, a cause who is competing with natural causes rather than working through them.

It is easy for this confusion to set in because in our everyday experience, when we attempt to explain the cause of something, we naturally rule out different competing causes. For example, if I throw a ball through a glass window, that naturally implies that my son was not the cause of the ball's flight. Either I caused the ball's unfortunate trajectory, or my son caused it. My son and I are competing causes. Unfortunately, many people view God's creative causality in much the same way. God is just one more possible cause like you or me or the force of gravity. Using this reasoning, if the trajectory of an object falling toward earth can be explained by the force of gravity, then God had nothing to do with it.

This is the framework behind many popular atheistic arguments bent on demonstrating the futility of religion and belief in God. These arguments all begin with the assumption that primitive human societies simply conjured up a belief in a higher power solely to explain the cause of events that they did not fully understand. For example, thunder was caused by the anger of the gods, disease was punishment from a god for injustice, and rain was caused by the proper ceremonial ritual dance that curried favor with the gods. Such beliefs have dissipated over time because as science has advanced, it has been able to offer compelling naturalistic accounts for many of these events, thereby removing God from the picture. The assumption is that as science continues to advance, eventually everything—from our origin as a species to our perception of morality to our religious practices—will all be explained fully by natural causes. Eventually there will be no more room for God (if room even still exists today) as he is increasingly outcompeted by natural processes. Science makes God increasingly irrelevant.

While this view is all too common, it depends upon a distinctly impoverished (and decidedly non-Catholic) understanding of God. God is not just one cause among many. Rather, God is, as St. Thomas Aquinas explained, the primary cause of everything that exists. St. Thomas distinguished primary causality, God causing all things simply to be (giving them existence), and secondary causality, the ability of created things like you and me to act in the world. Understanding this distinction is *the* key to developing a proper analysis of the relationship between evolution and creation.

Anytime something occurs in the universe, *both* primary and secondary causes are responsible; one type of causality does not subtract from or diminish the other because these are different types of causes that operate on different levels. The physicist Stephen Barr, founder of the Society of Catholic Scientists, often uses the play *Hamlet* as an analogy to illustrate this principle.[4] Near the end of the play, Laertes kills Hamlet by stabbing him with a poisoned sword. If one were to ask who killed Hamlet in the play, one could correctly assert that Laertes did. Yet, one could also claim that Shakespeare killed Hamlet inasmuch as

4. See Stephen Barr, *The Believing Scientist: Essays on Science and Religion* (Grand Rapids, MI: Eerdmans, 2016).

Shakespeare is responsible for creating the entire play and all the characters and actions within it. Thus, the characters (the secondary causes) can only do what Shakespeare (the primary cause) has allowed them to do. Shakespeare and Laertes do not compete as mutually exclusive causes of Hamlet's death, but rather are both causes of Hamlet's death, albeit on different levels.

Of course, when trying to explain God's primary causality, no analogy is perfect. However, the Hamlet analogy illustrates how different types of causes can operate on distinctly different levels. In a similar but not identical manner, God causes the existence of things such that they exist with certain causal abilities, abilities that can be used to do things in the world around us. These secondary causes are the type of causes that science can examine through its methods. Yet an understanding of how natural processes may have caused the emergence of bipedal hominins should not undermine God's causative role in the process. According to the theologian John O'Callaghan, the distinction between primary and secondary causation "forces us to recognize that the natural causal factors operative in evolution are only operative insofar as God renders them so by creating and sustaining them."[5]

If the creatures of the Earth around us evolved via natural causes, they did so only because God, as the primary cause of all that exists, allowed and sustained the proper secondary causes such that they could operate at every moment during the past four billion years. A proper understanding of this distinction creates the intellectual space to examine freely the evolutionary process without diminishing the role of God as Creator. It also allows us to examine the Genesis creation text in a more fruitful manner.

5. John O'Callaghan, "Evolution and Catholic Faith," in *Darwin in the Twenty-First Century: Nature, Humanity, and God*, ed. P. Sloan, G. McKenny, and K. Eggleson (South Bend, IN: University of Notre Dame Press, 2015), 280.

THE GENESIS TEXT

"In the beginning when God created the heavens and the earth . . ." (Gen. 1:1). These first words of the Bible, which set the stage for the creation stories that follow—not to mention the entire story of salvation history—make clear that God is the primary cause of the existence of every created thing in the universe. However, there is more to the Catholic understanding of creation and to the Genesis creation accounts than this.

The Genesis text contains two specific creation narratives. The first narrative, which spans Genesis 1 and the first lines of Genesis 2, gives the six-day creation account during which God progressively produces light, the heavens, the earth, the sun and moon, the living creatures, and ultimately mankind in the span of six days. In this account, mankind is created, both male and female, in the image of God, and then on the seventh day God rests. The second account, found in Genesis 2, gives no account of the formation of the earth or the heavens, but it does describe the formation of man from the union of the dust of the ground and the breath of God. The first man, Adam, is given dominion over the animals and names them. After failing to find a suitable companion for Adam among the animals, God then places him under a deep sleep while he forms the first woman, Eve, from Adam's rib.

What should a Catholic make of these texts regarding the creation of both the universe and mankind, particularly in relation to the science of evolution? There are those who feel that these two stories should be viewed as descriptions of actual historical events (putting aside the inherent discrepancies in the two accounts). If this is the case, the creation accounts and the theory of evolution are necessarily in direct competition with each other.

However, when considering the Genesis text, it is critical to recognize that the Church specifically rejects biblical fundamentalism, the position that the entire Bible corresponds exactly to historical fact. The Church realizes that the Bible frequently uses figurative or poetic language to make salient points. Referring specifically to the Genesis creation account, the *Catechism* states, "Scripture presents the work of the Creator symbolically as a succession of six days of divine 'work,' concluded by the 'rest' of the seventh day."[6] Catholics then have no need to necessarily equate creation with a historical understanding of the events of the first chapters of Genesis. Even when it comes to the story of the fall, the *Catechism* points out that the text "uses figurative language, but affirms a primeval event."[7] In other words, we know our first parents turned from God and rejected his gift of original holiness, but the exact manner by which they did so is not described historically in Genesis.

In our efforts to elucidate the proper approach to the interpretation of the Genesis texts, we owe a great deal to the work of Joseph Ratzinger. In the 1980s, he published a series of homilies addressing this very issue. In this book, entitled *"In the Beginning . . .": A Catholic Understanding of the Story of Creation and the Fall*, he makes an important distinction between the images of the Genesis stories and the truth the images are meant to convey. He points out that the "form [of Genesis] would have been chosen from what was understandable at the time—from the images that surrounded the people who lived then. . . . Only the reality that shines through these images would be what was intended and what was truly enduring."[8] This is echoed in the *Catechism*, which states, "In order to discover the sacred authors' intention, the reader must take into account the conditions of their time and culture, the literary genres in use at that time, and the modes of feeling, speaking, and narrating then current. For the fact is that truth is differently presented and expressed in the

6. *CCC* 337.

7. *CCC* 390.

8. Joseph Ratzinger, *"In the Beginning . . .": A Catholic Understanding of the Story of Creation and the Fall* (Grand Rapids, MI: Eerdmans, 1995), 5.

various types of historical writing, in prophetical and poetic texts, and in other forms of expression."[9]

Pope Benedict's approach mirrors the guidance found in the Vatican II document *Dei Verbum*, which states that "to search out the intention of the sacred writers, attention should be given, among other things, to 'literary forms.' For truth is set forth and expressed differently in texts which are variously historical, prophetic, poetic, or of other forms of discourse. The interpreter must investigate what meaning the sacred writer intended to express and actually expressed in particular circumstances."[10] Similar guidance was given by both Pope Leo XIII in the nineteenth century and Pope Pius XII in the middle of the twentieth century.[11]

With this in mind, one has to examine the Genesis text in light of what the authors were attempting to convey, being certain not to confuse the symbols of Genesis with scientific and historical explanations. Rather than scientific descriptions of the founding of the universe, the creation accounts in Genesis are more accurately viewed as vehicles employed by the sacred authors to convey profound messages to God's people—for example, that God created the world freely out of nothing, that the created order is good, that mankind was created in the *imago dei*, etc.

Reading Genesis in this manner does not undermine or trivialize the meaning of these early chapters of the Bible. In fact, the Catholic Church views the first chapters of Genesis as a description of creation that is alive with truth, meaning, and significance, one that sets the foundation for salvation history. As Christ frequently did with his use of parables, an illustrative story is often the most effective way to convey important truths. In the case of Genesis, the divinely inspired authors used these

9. *CCC* 110.

10. Dei Verbum 12, in *The Word on Fire Vatican II Collection: The Constitutions*, ed. Matthew Levering (Park Ridge, IL: Word on Fire, 2021), 28.

11. This was discussed in detail in chapter 3.

texts, which were filled with images familiar to their audiences, to convey specific messages. This is what Pope Benedict meant when he stated that the authors of Genesis would have "chosen from what was understandable at the time—from the images which surrounded the people who lived then."[12] The early chapters of Genesis are not a historical description of the founding of the world, but rather, as the *Catechism* states, "express in their solemn language *the truths of creation*—its origin and its end in God, its order and goodness, the vocation of man, and finally the drama of sin and the hope of salvation."[13]

ARE WE UNDERMINING THE BIBLE?

While it should be clear that Catholic exegetical principles do not require a historical reading of Genesis, there are many Christians who are convinced that "compromising" on the historical accuracy of the Genesis creation accounts will undermine the foundation of the whole of Christian theology. First, there is the concern that the need to reinterpret Genesis in light of modern science would undermine the authority of the Church, as the Church would have to backtrack from earlier interpretations of Genesis. For example, St. Augustine, like most of the Church Fathers, *did* believe that Adam was formed from the dust of the ground and that Eve was formed from the rib of Adam. Even the 1909 Pontifical Biblical Commission document "Concerning the Historical Character of the First Three Chapters of Genesis," commissioned and approved by Pope Pius X, stated that Catholics should affirm that these were historical events. If, because of the evidence for the evolutionary connection of man's physical form to other primates, we now see these as figurative descriptions of events, does this mean our Catholic beliefs have been diminished by science?

Such a view would be an overstatement. Rather, it seems

12. Ratzinger, *"In the Beginning,"* 5.
13. *CCC* 289 (emphasis added).

more proper to state that our understanding of Scripture and the truths of the faith have *developed* (not diminished) because of our understanding of the science. While many Catholics down through the centuries thought that man was specifically formed from the dust of the ground and/or that Eve was formed from Adam's rib, these ideas have never been pronounced as definitive teachings of the Church. Even Pope Pius X made clear that Catholics were "bound by the duty of conscience to submit to the decisions of the Biblical Pontifical Commission"[14] only regarding matters of doctrine. Now, certainly these age-old historical understandings of the text are related to doctrinal teachings of the Church, such as the belief that man is a unity of body and soul and the belief that man's soul is created immediately by God. However, an evolutionary understanding of the origin of man does not necessarily undermine these doctrinal teachings; it simply alters our understanding of how these have played out in history. In this sense, our Catholic beliefs are not diminished by interpretations inspired by our newfound scientific knowledge. In fact, one could argue our doctrinal beliefs have the opportunity to develop a richer, more profound depth in light of this new knowledge.

An additional and more serious concern is that if one claims the Genesis accounts are not accurate historical descriptions, then what is to prevent people from claiming other key parts of Scripture are not accurate historical descriptions of events? Maybe the account of Jesus' miracle of the loaves and fishes is not meant to describe an actual miracle. Or even more concerning, is it possible that something as central to Christian theology as the account of Jesus' death and Resurrection is not an actual historical event? The fear is that once the "skeptical" foot is allowed in the door,

14. Pius X, *Praestantia Scripturae*, motu proprio, November 18, 1907, in *The Sources of Catholic Dogma*, ed. Henry Denzinger and Karl Rahner, trans. R.J. Deferrari (St. Louis, MO: Herder, 1954), 543.

the whole fabric of Christianity, as revealed through Scripture, is prone to unravel.

This is a legitimate concern, particularly if Christians are arbitrarily deciding based upon the whims of modern science and culture which passages to take as historically accurate descriptions of events. In this case, there would be no clear manner by which to interpret Scripture, and each person would be left to his or her own devices to interpret any portion of the Bible as he or she saw fit. The Catholic Church, however, does not leave us to the ambiguous wilds of individual scriptural interpretation. The Catholic Church has a robust method for determining and defending which portions of the Bible should be read as historical descriptions and which should be read in other fashions. Thus, Catholics have a definitive body to turn to for guidance in this exercise.

One of the key principles behind proper biblical exegesis for Catholics is the understanding that any book or passage of the Bible must be read in the context of the whole. As described in the *Catechism*, "Scripture is a unity by reason of the unity of God's plan, of which Christ Jesus is the center and heart."[15] As a result, a narrow focus on a specific section of the Bible in isolation often leads to confusion. Ratzinger makes this point, stating that "every individual part derives its meaning from the whole, and the whole derives its meaning from its end—from Christ. Hence, we only interpret an individual text theologically correctly (as the Fathers of the Church recognized and as the faith of the Church in every age has recognized) when we see it as a way that is leading us ever forward, when we see in the text where this way is tending and what its inner direction is."[16]

Therefore, to understand Genesis properly and assess its true meaning—historical or otherwise—one has to look at it through the lens of the entire Bible. One must also read it in relation to

15. *CCC* 112.

16. Ratzinger, *"In the Beginning,"* 9–10.

Christ, with the aid of the apostolic tradition and the guidance of the Church, which "exercises the divinely conferred commission and ministry of watching over and interpreting the Word of God."[17] As a result, a Catholic is not left to his or her own devices to pick and choose which passages to take as historically accurate based upon a personal prejudice. Rather, he has a whole apostolic and teaching tradition upon which to rely.

Likewise, the Church ensures that any interpretation of a specific section of the Bible does not do violence to the unity of the entire Bible. For example, if one were to take Jesus' death and Resurrection as a symbolic description and not a historically accurate event, it throws the entire text of the Bible into disarray. Paul's references to Jesus' death and Resurrection no longer make any sense, nor does Jesus' own foretelling of his Passion, nor do the prophecies of Isaiah. In fact, the whole of the Bible points to the historical accuracy of Jesus' death and Resurrection. It is something that cannot be dismissed without undermining the whole text. The same cannot be said of the historical accuracy of the creation stories of Genesis, which is why the *Catechism* can state that "Scripture presents the work of the Creator [in Genesis] symbolically."[18]

In fact, when it comes to the creation accounts, a purely historical approach would do the Scriptures a great disservice by obfuscating rather than revealing underlying truths. For example, it would improperly anthropomorphize God to assume that he reached in and removed a rib from Adam or that he was walking in the garden or that he physically breathed into Adam's nostrils. In addition, the Bible is devoid of any references or passages that depend upon a historical interpretation of the creation stories in Genesis, so such a reading is not necessary to maintain the unity of Scripture. Even references to Christ as the new Adam do not rest upon a literal historical interpretation of Adam, but rather

17. *CCC* 119.
18. *CCC* 337.

hinge on the understanding that our first parents, whoever they may have been, turned from God and his plan for humanity.

What the Scriptures do contain are many references, usually couched in beautiful and poetic imagery, to God as the Creator of the heavens and earth. The book of Wisdom states, "For from the greatness and beauty of created things comes a corresponding perception of their Creator" (Wis. 13:5). Likewise, Psalm 33 reads, "By the word of the LORD the heavens were made, and all their host by the breath of his mouth. He gathered the waters of the sea as in a bottle; he put the deeps in storehouses. Let all the earth fear the LORD; let all the inhabitants of the world stand in awe of him. For he spoke, and it came to be; he commanded, and it stood firm" (Ps. 33:6–9).

While these passages and many similar ones refer to God as the Creator and Author of the universe, they do not hinge upon the historical accuracy of the Genesis creation accounts, nor do they provide scientifically accurate creation accounts themselves. If the universe were 13.8 billion years old rather than a few thousand, it would not undermine the profound truths to which any of these passages speak.

SCIENCE AND A NAÏVE LITERALISM

The need to understand both the intent of the inspired author and the literary style the author employed is essential for proper biblical exegesis. Uncritical literal/historical interpretations are dangerous because they set up conflicts where none need exist, as in the infamous case of Galileo. Galileo found himself at odds with those who felt his advocacy of the heliocentric theory conflicted with, among other passages, a literal interpretation of what the Psalmist wrote in Psalm 93:1: "He has established the world; it shall never be moved." Galileo was ultimately prosecuted and convicted of "vehement suspicion of heresy," in large part because he argued that the heliocentric system required one to reinterpret certain biblical passages, such as Psalm 93.

Today, there are few people (if any) who take the Psalmist literally and hold that the Earth is stationary. Christians rightly realize that the author was not *intending* to teach us about astronomy and the movements of the heavenly bodies in Psalm 93. Rather, the author was employing poetic language, which was influenced by the ancient understanding of the movement of the Earth and the heavens, to illustrate the point that God is the Creator of the world and that there is an order and purpose to his creation.

Yet, when it comes to Genesis, there are people (some Catholics included) who believe that the author of the sacred text *is* trying to teach us about cosmology, the origin and development of the universe, or the details of the physical creation of man. They believe that the author of the sacred text *is* trying to teach us that God made the fish on the fourth day and the land animals on the fifth day. Holding these views, against the intent of the sacred author, sets up a needless and potentially scandalous conflict between science and faith, much as the view about the stationary Earth did in Galileo's time. Because theologians and philosophers had interpreted passages of Scripture in light of the provisional account of a stationary Earth that was prevalent at the time, they had a difficult time reinterpreting these passages to accommodate a heliocentric view even as the data accumulated in support of that theory.

It is just this type of conflict that St. Augustine, a Church Father who did not hold to a strict historical interpretation of the Genesis creation accounts, was particularly concerned with avoiding. In his *The Literal Meaning of Genesis*, Augustine made clear that, when possible, one should be willing to reinterpret passages of Scripture in light of new findings:

> In matters that are obscure and beyond our vision, even

> those we find in Holy Scripture, different interpretations are sometimes possible without prejudice to the faith we have received. In this case, we should not rush in impetuously and adopt a position on one side with such commitment that, if further progress in the search of truth should undermine this position, we too should fall with it. That would be to contend, not for the teaching of Holy Scripture, but for our own teaching. We would wish its teaching to conform to ours, when we ought to wish ours to conform to that of Holy Scripture.[19]

While Augustine's rationale for interpreting Scripture considering new human knowledge was to conform ourselves to the wisdom of the Bible, such an approach, in which scientific wisdom and the interpretation of Scripture were synchronized, also had apologetic benefits. Augustine made the following point: "Even a non-Christian knows something about the earth, the heavens, and the other elements of the world. . . . Now it is a disgraceful and dangerous thing for an infidel to hear a Christian, presumably giving the meaning of Holy Scripture, talking nonsense on these topics; and we should take all means to prevent such an embarrassing situation, in which people show up vast ignorance in a Christian and laugh it to scorn."[20]

This does not necessarily mean that St. Augustine would have supported evolutionary theory, or that he would have advocated for an evolutionary worldview, but it does show he had a deep respect for honest scientific inquiry. More importantly, his interpretation of Genesis is such that it can possibly accommodate an evolutionary understanding of creation. In trying to reconcile the two creation accounts, Augustine argued that God created the universe in an instant with an inherent potency to produce

19. Augustine, *The Literal Meaning of Genesis* 1.18.37, trans. John Taylor (New York: Newman, 1982).

20. *Literal Meaning of Genesis* 1.19.

life forms over the course of time. According to the Protestant theologian Alister McGrath,

> Augustine's basic argument is that God created the world complete with a series of dormant multiple potencies, which were actualized in the future through divine providence. . . . God must be thought of as creating in that very first moment the potencies for all the kinds of living things that would come later, including humanity. . . . Augustine seems to have conceived of them as dormant "virtual" entities, enabling the natural world to emerge in its own way and in its own time. . . . This process of development, Augustine declares, is governed by fundamental laws, which reflect the will of the Creator.[21]

This gradual unfolding of what Augustine called the "rationes seminales," the "seed-like" potentials for each living form, is remarkably well suited to an evolutionary understanding of creation. To be clear, Augustine thought each form emerged from the potencies within the created order, and he did not advocate for one life form emerging from another. However, his belief that natural causes could allow for new life forms to emerge from the underlying created order is consistent with evolutionary ideas. As the Protestant theologian Gavin Ortlund has pointed out, "It should, at least, discourage us from dismissing evolutionary processes of creation *a priori*, as though God must always create quickly and directly, rather than slowly and indirectly, for it to be really a work worthy of divine power."[22] While there are ongoing, unresolved debates regarding how well his ideas fit with evolutionary theory, the key point here is that he is clearly not

21. Alister McGrath, *A Fine-tuned Universe: The Quest for God in Science and Theology* (Louisville: Westminster John Knox, 2009), 103.

22. Gavin Ortlund, *Retrieving Augustine's Doctrine of Creation: Ancient Wisdom for Current Controversy* (Westmont, IL: InterVarsity, 2020), 193.

advocating for a strictly historical interpretation of the Genesis creation accounts.

Augustine was not alone among the Church Fathers in this respect. Origen, St. Basil, and St. John Chrysostom all emphasized that the audience the Scriptures were written for should be taken into account when interpreting the creation accounts of Genesis. Origen in particular stressed the problems that occurred with reading the text literally:

> For who that has understanding will suppose that the first, and second, and third day, and the evening and the morning, existed without a sun, and moon, and stars? And that the first day was, as it were, also without a sky? And who is so foolish as to suppose that God, after the manner of a husbandman, planted a paradise in Eden, towards the east, and placed in it a tree of life, visible and palpable, so that one tasting of the fruit by the bodily teeth obtained life? And again, that one was a partaker of good and evil by masticating what was taken from the tree? And if God is said to walk in the paradise in the evening, and Adam to hide himself under a tree, I do not suppose that anyone doubts that these things figuratively indicate certain mysteries, the history having taken place in appearance, and not literally.[23]

Again, this does not imply that any of these Church Fathers were evolutionists. Such ideas were still unknown to them. The point to be made here is that a nonhistorical treatment of the creation accounts of Genesis is not a novel concept introduced by the Church to deal with evolutionary theory. It is not some rearguard action implemented to confront a "problematic" scientific theory. Rather, it is a principle that predates Darwin by more

23. Origen of Alexandria, *On First Principles* 4.1.16, trans. Frederick Crombie, in Ante-Nicene Fathers, vol. 4, ed. Alexander Roberts, James Donaldson, and A. Cleveland Coxe (Buffalo, NY: Christian Literature, 1885), newadvent.org.

than a millennium. It is the outgrowth of a biblical exegetical principle that can be traced to the Church Fathers, who correctly recognized that 1) different portions of the Bible were written in different literary styles and 2) all portions should be viewed within the context of the style in which they were written.

Given the thoughtful and scholarly manner by which the Church interprets Scripture, the Church has no reason to be afraid of what science may reveal about creation. As stated in the *Catechism*, "Methodical research in all branches of knowledge, provided it is carried out in a truly scientific manner and does not override moral laws, can never conflict with the faith, because the things of the world and the things of faith derive from the same God."[24] Therefore, the Church is confident that the truths of Scripture, elucidated through a careful and competent reading of the Genesis creation accounts, will be compatible with the truths revealed about our material universe, elucidated through careful and competent scientific investigation.

THE RICHES OF THE GENESIS TEXT

What then are the key truths to be taken from the Genesis creation accounts, if they are not necessarily meant to convey historical descriptions of actual events? One key truth revealed by the text, which has been discussed above, is that God brought the world into existence (it had a beginning) and its existence is and continues to be dependent upon his creative action. Besides this foundational principle, it is worth focusing on three additional truths that shine through the creation accounts, truths that are particularly relevant to the intersection of creation and evolution.[25]

To begin this process, it is best to start by looking at how the images used by the authors of Genesis would have been viewed

24. *CCC* 159.

25. Entire books have been written on this topic, and to plumb the depths of the text would go well beyond the scope of this book. Here it is worth focusing on a few key points that are relevant to the topic of evolution.

through the eyes of their primitive audience. This allows one to identify, as Ratzinger explained, "the reality that shines through these images."[26] This audience, living centuries before Christ, existed in an unstable, exposed, and likely transient population. In this environment, ancient man must have felt himself to be at the mercy of the gods, as storms, floods, famines, war, and disease would all strike without warning. Even though the scientific explanation of why his whole group fell violently ill with fever and vomiting was closed to him, he knew that there had to be a cause for such an event. In their desire to elucidate these causes, ancient peoples had developed all types of explanations. For the most part, these explanations involved the machinations of unpredictable and distant gods. These demiurges, gods, and rival powers all seemed to be using the world as their personal playground, making it bend to their capricious whims. At any moment, these forces could inflict pain and suffering, drought, or famine. All of creation seemed to be at the mercy of irrational forces, forces that must have left man with a foreboding sense of dread.

According to Ratzinger, viewing the text through the eyes of the ancients allows us to "appreciate the dramatic confrontation implicit in this biblical text, in which all these confused myths were rejected and the world was given its origin in God's Reason and in his Word."[27] In the simple yet profound statement "God saw everything that he had made, and indeed, it was very good" (Gen. 1:31), the biblical author claims that the Earth is not the province of impersonal gods and dueling spirits but the work of a divine and loving Creator who has instilled order on his creation. As the theologian Christopher Baglow has stated, "This goodness is both manifested in its order and explicitly declared by God.

26. Ratzinger, *"In the Beginning,"* 5.

27. Ratzinger, 13.

Seven times He sees and declares its goodness, a number in the Bible that symbolizes completion and perfection."[28]

This then is the first additional point to stress regarding the Genesis creation accounts: that God has created an ordered and intelligible world. As the *Catechism* states, "Because God creates through wisdom, his creation is ordered."[29] This is particularly evident in the first three days of creation, where the events described depict this order emerging from chaos. Darkness and light are put into an orderly relationship with each other, as day would follow night. In a similar fashion, the water of the earth is separated from that of the heavens. The earth is arranged such that dry land is separated from the sea. From a formless void a rational world arises, a world that promises to be intelligible, a world that is proclaimed good. This truth is repeated throughout Scripture. In Proverbs chapter 8, it is said that God "marked out the foundations of the earth" and "assigned to the sea its limit," and the Holy Spirit is referred to as his "master worker" (Prov. 8:29–30).

Furthermore, while creation is endowed with order, only humankind has been endowed with the gifts necessary to discern this order, given that humankind is the only earthly creature fashioned by God in the *imago dei*. Being made in the divine image refers to the fact that we are infused with the gift of reason such that we can discern the good and the order in creation. According to the *Catechism*, "Our human understanding, which shares in the light of the divine intellect, can understand what God tells us by means of his creation."[30] As Ratzinger has written, the Genesis story, which proclaims that man is formed in the image of God, is "the decisive enlightenment of history," as it "put human reason firmly on the primordial basis of God's creating Reason."[31]

28. Christopher Baglow, *Faith, Science, and Reason: Theology on the Cutting Edge*, 2nd ed. (Downers Grove, IL: Midwest Theological Forum, 2019), 82.

29. *CCC* 299.

30. *CCC* 299.

31. Ratzinger, *"In the Beginning,"* 14.

The fact that a loving God has made the world intelligible bodes well for scientific inquiry. It is a promise that scientific inquiry, if pursued diligently and "in a spirit of humility and respect before the Creator and his work,"[32] will uncover truths about the world that complement what God has revealed to us through Scripture. For our purposes, it means that an honest, competent investigation into evolution done through the proper use of human reason should lead us toward God rather than push us further down the path toward atheism. God has provided an order and rationality to creation, and it is an order that he has given us the resources to discern. We should not be afraid to take up this task.

The leads naturally to the second additional point revealed in the Genesis text that is relevant to the discussion of evolution: God created mankind deliberately, and mankind is "the summit of the Creator's work."[33] Genesis chapter 2 speaks of God forming man from the dirt and breathing life into him, while Genesis chapter 1 simply states that "God created humankind in his image, in the image of God he created them; male and female he created them" (Gen. 1:27). While the details of man's creation are unclear from the text, one thing is not: man's creation was an intentional act; it was no accident. Not only was it not an accident, but man's creation is upheld as the crowning act of creation, given that it is the last act in the text before God rests on the seventh day. As the *Catechism* states, man "is the only creature on earth that God has willed for its own sake."[34]

Many evolutionary biologists are fond of saying that man is just a late-arriving chance twig on the immense evolutionary tree. Mankind is just an unforeseen accident on evolution's twisting highway. For some scientists, this is enough to justify atheism; for others, it fosters a deistic view of God as an absentee landlord,

32. *CCC* 299.
33. *CCC* 343.
34. *CCC* 356.

one who got the ball rolling some years ago but has now retired to spend eternity playing golf. According to this deistic view, God would have been as surprised as anyone when large-brained upright primates started roaming the savannahs of Africa.

The Genesis 1 story counters such scenarios, arguing that we are the intended product of creation. It also sets us apart from the rest of creation, as man's unique nature is revealed clearly in the text. Unlike the rest of creation, we are at once physical (formed from the dust of the ground) and spiritual (God has breathed into man's nostrils the breath of life). In addition, man is given "dominion over the fish of the sea and over the birds of the air and over every living thing that moves upon the earth" (Gen. 1:28). This special relationship we have with the Creator and the rest of creation is reaffirmed throughout the Bible. The book of Wisdom states, "For God created us for incorruption, and made us in the image of his own eternity" (Wis. 2:23). Jesus reassured his disciples of their special worth, telling them, "So do not be afraid; you are of more value than many sparrows" (Matt. 10:31). The Scriptures make this point abundantly clear: we are all wanted children, planned from the beginning by our loving Father. No matter how our bodily form emerged in its present state, God knew when, where, and how we were coming.

The third and final relevant point that the text reveals is that all of creation points toward and is striving to reach its fruition in God. The *Catechism* declares that the universe was created "in a state of journeying, toward an ultimate perfection yet to be attained."[35] This point is made evident in the familiar structure of the Genesis story in which the process of creation, rather than occurring instantly, unfolds in a rhythmic fashion over seven days. Each day begins with the phrase "Then God said" and ends with "Evening came, and morning followed," giving the story a poetic refrain as it builds toward its climax.

At the onset of every new day, creation seems to be progressing

35. *CCC* 302.

toward something, or more precisely, toward Someone. It is no accident that the story reaches its completion on the seventh day, the sabbath day of rest set aside for worship of the Creator. It does so because this is precisely the object toward which all of creation is oriented. As Ratzinger stated, "Creation is designed in such a way that it is oriented to worship."[36] This theme, embedded in the structure of the Genesis story, is echoed throughout the Bible. In the book of Wisdom, the author warns about those who "were unable from the good things that are seen to know the one who exists" (Wis. 13:1). The Psalmist writes frequently of how the mountains, the sea, and other aspects of nature praise and point to God. In fact, the authors of the divine text repeatedly maintain that creation should turn our hearts toward him.

Despite this reality, many people are under the illusion that the deeper one studies nature, the further one gets from God. For Catholics, this notion should seem absurd. If God created the world in all its splendor, order, and beauty, then the study of it should orient us toward the one who fashioned it. Albert Einstein, while not a theist, had it correct when he remarked that his study of nature "revealed such a superior Reason that everything significant which has arisen out of human thought and arrangement is, in comparison with it, the merest empty reflection."[37] For Catholics, peering into creation should reveal to us a reflection of God's beauty and wisdom. This reflection should spur us onward toward worship of the real thing: it should orient us toward God.

WHAT THE CHURCH BELIEVES ABOUT CREATION

What has been discussed in this chapter is by no means meant to be a comprehensive analysis of the Genesis creation texts or the Church's understanding of creation. Rather, it is simply meant to

36. Ratzinger, *"In the Beginning,"* 27.

37. Albert Einstein, *Mein Weltbild*, ed. Carl Seelig (Stuttgart-Zurich-Vienna: Europa, 1953), 21.

lay the foundation for a robust and fruitful engagement between Catholic theology and evolutionary theory. A proper understanding of creation as a metaphysical principle rather than merely as an event in time, a correct understanding of God's primary causality in relation to the created world, and a recognition of the manner in which the Church has interpreted Genesis are all essential for this task.

In the examination of the Genesis text, four key points that are related to the discussion of evolution have been highlighted. In summary, they are as follows:

- God is the Creator of the universe. All that exists owes its very being (at each and every moment) to his creative and sustaining power. (*CCC* 296, 338)
- God has created an ordered world that mankind has been equipped with the rational power to comprehend. (*CCC* 299)
- Man, who is at once a physical and spiritual being, was created deliberately by a loving Creator who has placed him at the summit of his creation. (*CCC* 342, 343)
- All of creation points toward and reflects the majesty of God. A study of God's creation should bring us closer to him, drawing us into worship. (*CCC* 339, 347)

Of course, these are not the only truths that emerge from the Genesis text (others will be discussed in later chapters, particularly in relation to original sin), but they are the ones most relevant to a discussion of evolution in general. These are the key truths, revealed to us through Scripture, the doctrinal teachings of the Church, and our proper use of human reason, that will help illuminate our examination of evolution.

With this in mind, the next chapter will explore the science of evolution and how the science might be integrated with a

Catholic understanding of creation. What will hopefully become clear is that one does not have to twist the science of evolution to force-fit it into a Catholic worldview. Certainly, there exist versions of evolutionary theory that are in direct conflict with the views of the Church on creation. Yet when one examines these "theories," it becomes apparent that they have infused their reading of the science of evolution with bad philosophical assumptions, thereby obscuring the scientific truths of evolution. Separating the details of evolutionary science from this philosophical baggage is *essential* if one is to cultivate a fruitful dialogue between evolution and creation. Once this is done, Catholics have nothing to fear from an honest investigation into the *science* of evolution. Such an investigation is the objective of the next chapter.

5

The Science of Evolution

There is nothing God does not wish to be investigated and understood by reason.[1]

—Tertullian

It is often said that there are two topics that one should avoid at all costs during polite conversation: politics and religion. Given the visceral reaction that it often elicits, perhaps evolution should be added to this list. Since evolution touches on so many fundamental questions—where did we come from, who are we, what is our ultimate purpose—it is not surprising that it can become such a contentious subject. To make matters even worse, conversations regarding evolution can quickly hit an impasse given that people are seldom in agreement regarding how they are using the term.

The oft-quoted conversation in *Through the Looking Glass* between Humpty Dumpty and Alice illustrates this linguistic problem:

> **Humpty Dumpty**: When I use a word, it means just what I choose it to mean—neither more nor less.
>
> **Alice**: The question is, whether you *can* make words mean so many different things.[2]

The truth is, with evolution, people do make the word mean so many different things. Some people use the word to describe

1. Tertullian, *De paenitentia* 1.2, in *Treatises on Penance: On Penitence and Purity*, trans. William P. LeSaint (Mahwah, NJ: Paulist, 1959), 14.

2. Lewis Carroll, *"Alice's Adventures in Wonderland"* and *"Through the Looking Glass,"* edited with introduction and notes by Peter Hunt (Oxford: Oxford University Press, 2009), 190.

an atheistic materialistic process that removes the need for a Creator. Others use it to simply describe organisms changing over time. Others use it as a comprehensive scientific explanation for the natural world. Still others use it to describe a process that reveals God's creative power. This can quickly lead to confusion and derail any meaningful discussion.

The first step toward deciphering how someone is using the term evolution is to recognize the distinction between *scientific* theories of evolution and the *philosophical* claims that are so often associated with them. For the term evolution to be used scientifically, it must be restricted to making claims regarding the material world, claims that can be empirically supported via experimentation and observation. If a "theory" of evolution makes claims regarding the existence of God or the meaning of life, it has stepped outside the confines of modern empirical science and has entered the realm of philosophy. This can lead to a good deal of confusion regarding what are and what aren't scientific conclusions.

To illustrate how philosophical claims can become intertwined with scientific theories of evolution, it is useful to examine how Darwin's theory of evolution is applied to the origin of humans. Darwin's theory makes the scientific claim that the process of natural selection can account for the emergence of the diversity of species that have appeared on our planet over the past four billion years. The theory would claim that even the physical form of man emerged via this natural process. Such a claim is scientific in nature, as it is making claims about the physical world (man's physical form) that can be supported by experimentation and observation. Darwin himself, though, took this scientific claim one step further. In his book *Descent of Man*, Darwin not only argued that man's physical form emerged via natural selection (a scientific conclusion), but he made it clear that he thought man, with all his myriad intellectual and mental abilities, could be explained entirely by his theory (a philosophical conclusion). In doing so, Darwin staked out a clear philosophical position—namely, that man is nothing more than a material being, a being that can be explained by modern science *in his entirety*. Because Darwin seemed to think that man was nothing more than a material being, this position is often assumed to be part of Darwin's *scientific* theory of evolution. But such a position cannot be part of Darwin's scientific theory because it is not a scientific claim. There is no experiment or observation that one could conceivably make to scientifically demonstrate that 1) man is a unity of body and soul or 2) man is merely a conglomerate of matter. The question of whether man has an immaterial aspect to his person is not one science is equipped to ascertain.

Stripping away this type of philosophical baggage from evolutionary theory is a constant battle, as those on both sides of the evolution debate continually blur the distinction between the science of evolution and the philosophical claims they believe flow from the science of evolution. In fact, everything from atheism to fascism to eugenics to theism have all been "associated" with evolutionary theory. However, none of these flow by necessity from the science. These are philosophical positions that people have appended to evolutionary theory for reasons that go well beyond the scientific data.

Given this problem, it is critical to understand what the science of evolution actually demonstrates. In particular, what scientific claims can one make regarding evolution given the evidence we have at our disposal? Does the evidence support the scientific position that mankind shares a common ancestry with other primates? Does the evidence support the idea that the physical forms of organisms emerged over long periods of time through the operation of material processes? It is the purpose of this chapter to give a brief overview of these scientific questions.

EVOLUTION AND THE TENTATIVE NATURE OF SCIENCE

Even after the philosophical baggage is stripped away, the term evolution can still take on many different *scientific* meanings. As a result, it is necessary to qualify the term when examining the "science of evolution," and for our purposes it is useful to distinguish three tiers or levels of evolution. The most foundational level or basic level is *historical evolution*, which is the claim that the Earth is over four billion years old and the species that have inhabited the planet have varied considerably over this time span. For example, the species that existed 400 million years ago were quite distinct from the species that existed 50 million years ago.

If this foundational level is established, one can then move up to the second level of evolution, *common descent.* Common descent posits that all the species that have existed over the life of our planet are related and have descended from a common ancestor or a group of common ancestors. This would imply that humans and chimps descended from a common ancestor population that diverged and eventually gave rise to modern humans on one branch and modern chimps on another. The validity of universal common descent is dependent on historical evolution, as it assumes that the Earth is old and the type and distribution of species upon the Earth has varied over time, but it makes a claim that goes beyond it—namely, that all these species are related via common descent.

The final level, which is dependent upon the establishment of the underlying two levels, is *naturalistic evolution*. This is the level of evolution that claims to provide an explanation for common descent. Naturalistic evolution posits that natural processes, such as natural selection, can drive changes in species over time such that the pattern of common descent emerges. It posits natural explanations for the entire physical evolutionary process, a process that has given rise to the historical pattern of common descent seen in the fossil record. While these three levels are interrelated, each one makes distinct scientific claims, each can be supported by distinct pieces of scientific evidence, and each is associated with distinct theological and philosophical implications.

What evidence do we have for each of these levels? Before diving into this question, it is helpful to clarify what one can reasonably expect modern empirical science to demonstrate regarding the different levels of evolution so that we don't expect too much (or too little) from the science. Modern science is an extremely useful tool for gaining knowledge of the material world. Scientists do this by observing the physical aspects of the world and by performing scientific experiments. Based upon the factual evidence that they gather—such and such a fossil was found in this rock layer, etc.—scientists can then develop theories or explanations about the material world to explain these facts. These theories are interpretations of the evidence, and good theories are 1) able to explain a large amount of the evidence at hand and 2) supported by most of the available evidence.

What scientific theories do *not* provide is absolute certainty. This is particularly the case when it comes to evolution, which is largely a historical science. As a result, developing a *scientific proof* that specific natural processes drove an ancient evolutionary transition is not possible. For example, when investigating the transition from theropod dinosaurs to birds, one cannot assemble a group of theropod dinosaurs in a lab, vary the conditions, and then wait a few million years to see what happens to fall out. We do not have the ability to rerun the evolutionary tape and see exactly what might have occurred at the major evolutionary transitions that are evident in the fossil record. This makes it difficult to ascertain with certainty the key factors that may have influenced specific transitions and the exact role different natural processes such as natural selection played in the transition.

While it is important to acknowledge these inherent difficulties, this does not imply that evolutionary biologists who study the theropod dinosaur to bird transition are on a fool's errand. Rather, it just means that they have undertaken a difficult task, the task of assembling a compelling range of circumstantial evidence to make the case that this transition actually occurred via natural processes. The situation they find themselves in is analogous to a lawyer making a case in a murder trial. Assuming there was no surveillance camera filming the crime, the lawyer is left to assemble the evidence—items left at the crime scene, the motivation of the subject, the timing of the crime, etc.—into a coherent case that can establish the likelihood that the suspect did in fact commit the crime. The lawyer can also supplement this historical data with lab data on blood or hair samples (much like evolutionary scientists can use radiometric dating techniques and DNA comparisons to bolster their case). However, even if the lawyer is able to assemble a compelling range of evidence, absolute certainty is ruled out. The suspect could have been framed or new evidence could emerge at a later date that could change the outcome of the trial. This is why legal cases are not held to the standard of absolute certainty, but rather to the standard of beyond a reasonable doubt.

Evolutionary scientists face a similar albeit more formidable task when it comes to establishing things like the theropod dinosaur to bird transition. To make their case, paleontologists are left to scour the fossil record, a "crime scene" that in this case has sustained roughly 150 million years of geological alterations. In addition, developmental biologists can perform lab experiments on bird development to posit possible ways that birds could have evolved from dinosaurs. Likewise, evolutionary geneticists can compare DNA sequences of living and recently extinct species to assemble possible evolutionary histories of birds and reptiles. Combining all of these various lines of research, one can then attempt to construct a coherent narrative of natural history.

Remarkably, despite all the difficulties described above, this is what has occurred with evolutionary studies. Independent lines of research in multiple disciplines have been successful in establishing a solid case for the evolutionary relatedness of all living creatures—that is, common descent. This remarkable convergence of different disciplines was what John Paul II referred to in his address to the Pontifical Academy of Sciences in 1996 when he stated, "The convergence, neither sought nor provoked, of the results of work that was conducted independently is in itself a significant argument in favor of this theory [of evolution]."[3] Of course, there is always new evidence that can emerge to alter the theory and require its revision and rethinking. In fact, there are a number of developments in evolutionary science that have occurred over the past few decades that have led many scientists to suggest a significant revision to the standard Darwinian understanding of the processes driving evolution. While this work has not necessarily displaced the role of natural selection, it has led to an appreciation of other processes that work alongside natural selection. The synthesis of this work is referred to as the extended evolutionary synthesis and is discussed in more detail below.

The key point, though, is that science does not provide certainty, and evolutionary science, given its historical nature, entails even less certainty. As the philosopher Nicholas Rescher points out, "If the future is anything like the past, if historical experience affords any sort of guidance in these matters, then we know that all of our scientific theses and theories at the present scientific frontier will ultimately require revision in some (presently altogether indiscernible) details."[4]

This does not imply that our science is necessarily incorrect, but merely that it is likely incomplete. There are many aspects of evolutionary history that currently remain masked behind the veil of time. Such is the reality of a science that relies upon interpretations of billion-year-old rocks and DNA comparisons that

3. John Paul II, "On Evolution" 4, Message to the Pontifical Academy of Sciences, October 22, 1996, in *Origins* 26, no. 25 (December 5, 1996): 415.

4. Nicholas Rescher, "The Problem of Future Knowledge," *Mind and Society* 11, no. 2 (2012): 149–163.

can only involve the small number of species that have survived to the present day (or very recent past).

To reiterate, this is not an attempt to dismiss evolutionary science in general or to dismiss any specific hypotheses, such as the evolutionary connection between theropod dinosaurs and birds. Rather, it is a cautionary tale to both evolutionary advocates and critics alike. If one criticizes and dismisses an evolutionary picture of natural history merely because it is incomplete or tentative, to be consistent one must dismiss all of science. If one thinks evolution should be able to demonstrate the sweep of evolutionary history in the same manner by which two plus two can be shown to equal four, one sets all of science up for failure. Instead, the proper question to ask is the following: Given the evidence we have now at our disposal, what scenario or evolutionary hypothesis best makes sense of the data? It is all about assembling our current best estimate with not only the understanding but the *expectation* that some revisions are coming down the road. With this caveat in mind, it's worth looking at the current evidence for the three different scientific levels of evolution that we have delineated.

HISTORICAL EVOLUTION

The initial and most foundational of the three levels of evolution to discuss is historical evolution. While the term historical evolution is not in popular use, we will use it here to indicate the position that 1) the Earth is roughly 4.5 billion years old and that 2) over this time the organisms that have inhabited the planet have changed drastically both in complexity and diversity. For example, sedimentary rocks that are one billion years old will hold a different set of organisms than those found in rocks that are one hundred million years old.

Overall, there is excellent scientific evidence to support historical evolution. In the case of the age of the Earth, strong evidence for a four-plus-billion-year-old Earth has been gleaned from the radiometric dating of rocks. This dating method consists of measuring the amount of radioactive decay that has occurred in a rock since its formation. Basically, the older a rock, the more radioactive decay that has occurred. One common isotope to use for dating very old volcanic rocks is Potassium-40, which decays to Argon-40 very slowly. In fact, if you started with a pile of Potassium-40, it would take 1.3 billion years for half of the pile to decay. When volcanic lava cools and crystalizes in rock, there is usually only trace amounts of Argon-40 in it. As the Potassium-40 decays over time to Argon-40, more Argon-40 becomes trapped in the rock. One can then measure the current amounts of both Potassium-40 and Argon-40 isotopes in the rock sample and obtain an accurate estimate of the age of the rock.

In practice, there are many complications that affect such measurements, including the possibility of isotopes leaking in and out of the rock, as well as the heating and distortion of the rock over time. When these events occur, rocks known to have formed very recently can be measured to be millions of years old by standard radiometric dating methods. However, if one takes multiple samples and measures different parent-daughter pairs from within the same rock, this type of contamination usually can be identified. Because of this, the method is quite reliable and has been used to establish the standard geological column, a column that spans four billion-plus years on Earth.

The other major piece of evidence for historical evolution is the stratification of fossils in the geological column. The geological column contains distinct rock layers that are ordered in a relatively uniform pattern. These layers have names such as Cambrian rocks, Devonian rocks, and Cretaceous rocks, with the name often corresponding to the geographical locale of a particularly robust example of that rock layer. While one does not find *all* the layers neatly stacked one on top of the other anywhere on the globe (one would not expect this, as rock layers do not form uniformly across the entire surface of the Earth), they are nearly always found in the proper temporal sequence. For example, older rock layers such as Cambrian rocks are found underneath younger rock layers such as Cretaceous rocks. In addition, each of these rock layers contains a unique composition of fossils, which correspond to the organisms that dwelt on the Earth at the time that rock layer formed.

In general, more recent layers display forms with increasing morphological complexity and ecological specialization as compared to older layers. For instance, the Cambrian rocks contain the first animal fossils with shells, while the more recent Devonian rocks lack most of these early animals but contain more complex animals such as the early tetrapods (four-limbed animals, such as amphibians, reptiles, and mammals) that the Cambrian lack. Likewise, the Cretaceous rocks, which are more recent than the Devonian, lack the early tetrapods but contain a diverse range of more advanced tetrapods, including dinosaurs and birds. It is remarkable that each layer contains uniquely identifiable groups of fossils and that the layers are ordered in a progressive temporal sequence.

Given the strong scientific evidence for historical evolution, how does one integrate this with Church teaching? As discussed in chapter 3, there is no reason for Catholics to take the creation story in Genesis as a description of natural history and argue 1) that the world is only a few thousand years old or 2) that all life forms were created in a matter of days. Subsequently, historical evolution, the theory that the Earth is old and there have been incremental changes in the appearance of life forms over billions of years, in no way conflicts with Catholic teaching.

In fact, many people have commented on the parallel between the account in Genesis of the progressive populating of the Earth with living creatures and what historical evolution has revealed. In the Genesis 1 creation account, plants emerge first (albeit land plants), followed by sea creatures, then birds. This is followed by land animals and then finally humans on the sixth "day." The details of this account don't quite fit with historical evolution—for example, land animals appear before birds—but the progressive populating of the Earth as described in Genesis 1 does find parallels with the progressive appearance of different life forms as indicated by the robust evidence for historical evolution. Regardless of these similarities, it is important to recognize that historical evolution does not make claims regarding *how* the various creatures appeared (or how they are related). It merely posits that they do appear in a specific order in the fossil record, a record that spans billions of years.

COMMON DESCENT, UNIVERSAL AND OTHERWISE

While the majority of Catholics do not have issues with historical evolution (or the notion of an old Earth), things tend to become more controversial when it comes to the theory of universal common descent. This is largely due to the fact that this has direct implications regarding the origin of man. As a result, accepting universal common descent can be a bit more problematic for religious believers.

Universal common descent refers to the theory that all life forms, including humans, are related by a common ancestry insofar as they have all descended from an initial common ancestor. The most iconic image of this theory is the universal tree of life, a tree in which all the major groups of living organisms are found on the branches, while the universal common ancestor occupies the trunk of the tree. (This is akin to a family tree in which the great-great-grandparents represent the trunk of the tree, with all the subsequent descendants branching out from the trunk.)

This universal understanding of common descent needs to be distinguished from *limited* versions of common descent that are held by some who oppose the notion of *universal* common descent. An example of limited common descent would be the following: all species of elephants descended from a common ancestor, and all wolves and dogs likewise share a common ancestor, but species as disparate as dogs and elephants are *not* united by common descent. Such narrow versions of common descent, which are even held by some Young Earth Creationists, are distinct from the conventional scientific theory of universal common descent, the position that *all* living and extinct life forms have descended from a common ancestor that existed in the distant past.

The nature of this common ancestor, often referred to as the last universal common ancestor (LUCA), and the time of its existence are the subject of significant debate. Some argue LUCA was a community of different single-celled species out of which emerged all present life forms, while others posit that it was a distinct single-celled species. Given the paucity of fossils from three-to-four-billion-year rocks and the inherent ambiguities associated with fossils this old, the nature of the universal common ancestor may be forever cloaked in mystery.

Despite this, there exists a good deal of circumstantial evidence that all life forms are indeed related via common descent. The most "universal" piece of evidence is the fact that all life forms use basically the same genetic code. The genetic code refers to the way a cell reads the information in the DNA and then translates this information into the formation of a protein for use in the cell. The DNA is written in the language of nucleotides, long stretches of four bases popularly referred to as As, Ts, Gs, and Cs. Proteins are constructed in a different language, the language of amino acids, of which there are twenty common biological ones. To make a protein, which is a stretch of amino acids linked together, the cell must use the information in the DNA to determine the order in which the amino acids should be assembled. The process by which this is done is called translation, and it works much in the same way as translating English into Spanish.

While a language dictionary can be used to translate the English word "white" into the Spanish word "blanco," the genetic code is used to translate the DNA sequence into a protein sequence. The genetic code can be thought of as the list of rules for doing this translating. For example, the DNA sequence AAA is translated to mean that the amino acid lysine should be put into the protein. If the DNA sequence GGA comes next, this is translated to mean that the amino acid glycine should follow lysine, and so on until the end of the protein.

The interesting thing is that, while the universal genetic code is quite robust, there are a number of other possible genetic codes that would work just as well as the one found in virtually all cells; in fact, some alternative versions might work even better.[5] Just like there are many different human languages, most of which function quite well, there could have been many different "languages" of the genetic code. The fact that nearly all organisms share this particular one, even though there are many that would have fit the bill, suggests that the code originated in a common ancestor and was inherited by all subsequent life forms via common descent.

While the genetic code is probably the best piece of evidence for universal common descent, there are other pieces of evidence as well. All cells use double-stranded DNA molecules to store information, they all are bound by similar types of lipid membranes, and they share many of the same basic pathways for producing and using energy in the form of ATP molecules and for performing other key life-sustaining reactions.

Additional compelling support for universal common descent can be found by looking at the evidence for common descent on the various branches of the evolutionary tree. While it would not fully establish universal common descent, evidence that all

5. There are a few organisms that have slight modifications to the genetic code, but even in these cases the vast majority of the code is conserved. See chapters 1 and 2 of Simon Conway Morris' *Life's Solution: Inevitable Humans in a Lonely Universe* (Cambridge: Cambridge University Press, 2003) for a more detailed overview of this topic.

primates are united by common descent or that all mammals are united by common descent would be consistent with and provide strong support for universal common descent.

One particularly strong piece of evidence supporting this type of local common descent is the existence of homologous synteny blocks, which are similar regions of DNA that are conserved in different species. One can think of these synteny blocks as genome jigsaw pieces that existed in a common ancestor but that are put together in different ways in their descendants due to various large-scale genome rearrangements[6] that have occurred over time.

When a single mammalian population splits off into two separate species, the DNA would change independently in each species over time as specific genome rearrangements arise in one species but not the other (and vice versa). Despite these changes, large sections of their chromosomes will maintain the same order of genes. While these sections may be rearranged as pieces of the chromosomes break and fuse or are inverted or duplicated, the order of the genes within these large blocks will be maintained.

For example, chromosome 2 in humans appears to have originated via the ancestral fusion of chromosomes 2a and 2b, which are found in other primates. As a result, the order of genes on chimp chromosomes 2a and 2b is nearly identical to the order found on human chromosome 2. Even species as distantly related as humans and mice share a huge array of synteny blocks. Given that there seems to be no functional reason the genes found in these synteny blocks should be in a similar order in two distinct species (the likelihood of this happening by chance is vanishingly small), the most parsimonious explanation is that they were

6. Such rearrangements involve the fusing of two separate chromosomes into one larger chromosome, the splitting of a chromosome into two separate chromosomes, and the inversion of a piece of a chromosome in which the orientation of the piece gets flipped. All of these types of large-scale genome rearrangements are known to occur quite frequently in organisms.

inherited in that specific order from a shared common ancestorial species.[7]

The existence of these synteny blocks is not the only evidence that humans and other animals are related via common descent. There is a variety of additional evidence supporting this. Take for instance the fact that humans, unlike most mammals, lack the ability to synthesize vitamin C. This vitamin is important for the synthesis of the protein collagen, the most abundant protein in the human body. Most mammals carry a functioning copy of a gene called the GLO gene, which is needed to synthesize vitamin C in the liver. Without this gene, the liver cannot produce vitamin C, and it must then be obtained from the diet.

While humans lack the ability to make vitamin C, they do not entirely lack the GLO gene. Rather, they have a copy of it that doesn't work. The human version has accumulated a number of mutations such that it doesn't function properly. Interestingly, humans are not the only species that carry the mutated copy of the GLO gene. It turns out that chimps, gorillas, and orangutans, the primates most closely related to us evolutionarily, also have the same mutated gene with many of the exact same mutations. On the other hand, more distantly related primates such as lemurs have a functioning GLO gene. The most logical explanation for this arrangement is that sometime during primate evolution, the GLO gene became defective in a common ancestor and this defective version was subsequently inherited by a group of related primates, including humans.[8]

There are many other examples of odd modifications in our DNA that we share with other primates. Processed pseudogenes, which are copies of genes that have inserted in specific locations in the genome and degenerated such that they are no longer

7. For an overview of the concept of synteny blocks, see P.Z. Myers, "Synteny: Inferring Ancestral Genomes," *Nature Education* 1, no. 1 (2008): 47, https://www.nature.com/scitable/topicpage/synteny-inferring-ancestral-genomes-44022/.

8. For an easy-to-read overview of this, see Kenneth Miller, *Only a Theory: Evolution and the Battle for America's Soul* (New York: Viking, 2008), 97–99.

used to make proteins, represent another great piece of evidence in support of common descent.[9] Based upon how much the sequence in a pseudogene has degenerated, one can determine the approximate age of the pseudogene (when it originated). When one compares the location and age of these pseudogenes in primates, one finds a pattern that is best explained by common descent. The most recent pseudogenes (ten to fifteen million years old) are only found in humans, chimps, and gorillas, while the older ones (thirty million years old) are found not only in these species but also in more distantly related ones, such as orangutans and capuchin monkeys. The odds of these pseudogenes appearing simultaneously by chance in the same location in all these genomes in a temporal pattern consistent with common descent is astronomical. Common descent remains the best explanation for this arrangement of pseudogenes, an arrangement that is not unique to primates but occurs in a wide swath of eukaryotic organisms (plants, animals, and fungi).

Even critics of the efficacy of natural selection, such as the Catholic biologist Michael Behe, recognize the strong evidence that humans and other primates share a common ancestry. Behe states that one finds "the same mistakes in the same pseudo-gene in the same positions of both human and chimp DNA. If a common ancestor first sustained the mutational mistakes and subsequently gave rise to those two modern species, that would very readily account for why both species have them now. It's hard to imagine how there could be stronger evidence for common ancestry of chimps and humans."[10]

Behe correctly recognizes that there is a large body of evidence supporting the position that humans and other primates

9. An example of pseudogenes that are found in the same location in humans and chimps can be found in Daniel Fairbanks and Peter Maughan, "Evolution of the *NANOG* pseudogene family in the human and chimpanzee genomes," *BMC Evolutionary Biology* 6, no. 12 (2006), https://doi.org/10.1186/1471-2148-6-12.

10. Michael Behe, *The Edge of Evolution: The Search for the Limits of Darwinism* (New York: Free Press, 2007), 71–72.

are related via common descent. As discussed in chapter 2, such a position does not pose an inherent problem for Catholics. Pope St. John Paul II and Pope Pius XII have both articulated that one can examine the evidence of evolution in terms of how it refers to the origin of the human physical form via common descent, as long as one does not lose sight of the fact that the human soul (and therefore the human person) is "immediately created by God."[11] As long as the theory of common descent is seeking to understand biological developments and relationships, it does not contradict the essential Church teaching that man, who is at once a physical and spiritual being, was created deliberately by a loving Creator.

While there exists strong evidence for universal common descent, it is important to note that demonstrating common descent does not necessarily demonstrate a physical process to explain it. The evidence for common descent may establish that humans and chimpanzees descended from a common ancestor, but it does not necessarily explain *how* that process occurred. In fact, there are those like Behe who accept common descent but argue that natural mechanisms alone (particularly natural selection) are inadequate to explain it.[12] However, most biologists disagree with Behe's assessment. They believe that there are adequate scientific explanations for the pattern of common descent. It is at this point that one arrives at the final tier of evolution, the level of naturalistic evolution.

NATURALISTIC EVOLUTION

When people think of evolution, they naturally tend to think of Darwin's theory of natural selection, as it is the most widely known natural process that has been proposed to drive

11. This terminology is used by both Pope St. John Paul II in his Message to the Pontifical Academy of Sciences, October 22, 1994, and Pope Pius XII in *Humani Generis*, August 12, 1950.

12. For an overview of Behe's position, see Michael Behe, *Darwin's Black Box: The Biochemical Challenge to Evolution* (New York: Free Press, 1996).

evolutionary change. While there are a variety of other natural processes that also appear to be involved at the level of naturalistic evolution, it is best to begin with an examination of Darwinian evolution. Darwinian evolution is usually defined as the theory that natural selection operating over long periods of time is the primary mechanism that can explain universal common descent. However, despite bearing Darwin's name, this definition of Darwinian evolution does not exactly correspond to the more nuanced scientific position regarding evolution that Darwin staked out in his writing.

In the decades following the initial publication of the *Origin of Species* in 1859, Darwin's various ideas were evaluated alongside other evolutionary ideas such as saltationism, the idea that evolution proceeded by large leaps, and Lamarckism, the idea that organisms could acquire characteristics during their lifetime that could then be passed down to future generations. In addition, different theories regarding how genetic inheritance operated gave rise to different notions regarding how evolution actually proceeded.[13] The work of the Augustinian monk Gregor Mendel, which was published in 1866, indicated that the units of inheritance (genes) seemed to be inherited as discrete stable entities that passed from parent to offspring. Unfortunately, this work was largely unknown during Darwin's lifetime. For his part, Darwin thought that inheritance was controlled by small particles he called "gemmules" that were produced by various cells in the body and then transmitted via the gametes to the offspring. These gemmules could be altered by the organism's environment, and they would blend with the gemmules inherited from the other parent such that a cross between a red and a white flower should normally produce a blended intermediate pink form. Mendel's work suggested otherwise, but because it was not rediscovered until the turn of the century, there were considerable debates

13. In the late nineteenth and early twentieth centuries, the mechanisms governing genetic inheritance were still unclear.

in the late 1800s and early 1900s regarding this question. The outcome of these debates, fueled by the rediscovery and duplication of Mendel's work, eventually paved the way for what became known as the modern synthesis, an evolutionary view that still holds considerable sway, even today.

The development of the modern synthesis in the mid-twentieth century represented an attempt to integrate a modern understanding of Mendelian genetics (genes being inherited as stable discrete units) with Darwin's notion of natural selection. In general, this version of Darwinian evolution posited the following: 1) natural selection is the process by which novel species emerge (it explains common descent), 2) new forms are produced via this process primarily by natural selection operating on heritable changes in genes, 3) this process occurs gradually, and 4) other natural processes do not play significant roles in evolution.

Although such a view is consistent with passages from Darwin's *Origin of Species*, Darwin's writing overall does not neatly fit within the narrow confines of the modern synthesis.[14] For starters, Darwin, while advocating for the primacy of natural selection, believed that other processes may play a role in evolutionary change. This included the possible inheritance of acquired characteristics: "I think there can be no doubt that use in our domestic animals has strengthened and enlarged certain parts . . . and that such modifications are inherited."[15] While this was dismissed as heretical by the founders of the modern synthesis, the recently discovered phenomenon of epigenetic inheritance, in which heritable changes in gene expression can be both acquired by the individual and passed on to offspring through the gametes, suggests that Darwin's more expansive view is the more scientifically accurate one.

In fact, ever since Darwin published his ideas in 1859, there

14. For a good overview of the complexities of Darwin's views and the limited framework of the modern synthesis, Stephen J. Gould's *The Structure of Evolutionary Theory* (Cambridge, MA: Belknap, 2002) is an excellent although daunting resource.

15. Charles Darwin, *The Origin of Species* (New York: Mentor, 1958), 133.

have been rigorous *scientific* debates regarding the relative importance that various natural processes play in driving evolutionary change. Besides Darwin, many of his contemporaries, as well as late nineteenth and early twentieth century biologists, posited that natural processes besides natural selection might play key roles in evolution. This debate has continued unabated to the present, as during the past few decades, a diverse group of researchers have championed a collection of additional natural processes that they argue play important roles in influencing the outcomes of the evolutionary process. This collection of processes has been termed the extended evolutionary synthesis (EES) and overlaps with many arguments made by researchers in the late nineteenth century.[16] As the biologist and philosopher Lynn Chiu wrote, "Understanding the history of evolutionary biology will show us that it has never been a single, static, or unified research field, but a dynamic constellation of concepts, assumptions, and practices. Not everyone agrees with the entire package."[17]

While Chiu is correct that there is significant scientific disagreement within the field of evolutionary biology regarding the relative importance of the various processes involved, there is near universal agreement that common descent can be explained by natural processes. To give the reader some idea of the processes involved in the EES, a few of the more prominent ones are listed and described below:

- Evo-Devo: Small changes in genes affecting early development can trigger rapid, large-scale, phenotypic change in organisms. For example, a small mutation in a gene involved

16. The extended evolutionary synthesis incorporates a variety of additional processes—such as symbiosis and cooperation between organisms (particularly at the micro-organismal level), developmental programs that bias the evolutionary process toward certain phenotypes, cultural learning, and epigenetic inheritance—as key drivers of evolution. These processes work in conjunction with natural selection. There is considerable variability among advocates of the EES regarding the relative importance of natural selection vs. these additional natural processes.

17. Lynn Chiu, *Extended Evolutionary Synthesis: A Review of the Latest Scientific Research* (West Conshohocken, PA: John Templeton Foundation, 2022), 85, https://doi.org/10.15868/socialsector.40950.

in limb development can produce an entirely new bone structure in the limb. This favors the notion that evolution can proceed both rapidly and gradually.[18]

- Epigenetic change: During an organism's lifetime, the DNA can be modified to turn on or off certain genes in response to certain environmental factors. These modifications can be inherited and affect the phenotype of the offspring in a directed manner.

- Phenotypic plasticity: The organism's genetic material interacts with the environment such that the phenotype of the organism can change with the environment even if there is no genetic change. A given genotype can produce a range of adult phenotypes based upon the specific environment.

- Convergence: Organismal forms are limited and directed by the order and constraints in the underlying physics and chemistry. For example, the physics of distributing fluid from a central location (the heart) dictates the use of a fractal branching pattern, a pattern that has been repeated in numerous organisms for numerous purposes. Convergence implies that the same forms are repeatedly found by the evolutionary process because of the order in the physics and chemistry of our universe.[19]

- Symbiosis and DNA Exchange: Organisms cooperate to create communities and ecosystems that can evolve in novel ways. This also includes the swapping of DNA between various microbial organisms, a process that can produce novel genetic variants with new and surprising functions.

This is by no means an exhaustive list. Rather, it is meant to

18. For a good overview of Evo-Devo theory, Sean Carroll's *Endless Forms Most Beautiful: The New Science of Evo Devo* (New York: W.W. Norton, 2006) is a good resource.

19. For the seminal overview of this topic, see Simon Conway Morris' *Life's Solution: Inevitable Humans in a Lonely Universe* (Cambridge: Cambridge University Press, 2003).

give the reader a glimpse of the ongoing research and discussions regarding various potential evolutionary processes.[20]

While the EES has made significant inroads within the scientific community, natural selection still retains a place of prominence in evolutionary studies. Given this, it is worth discussing the specifics of natural selection. Contrary to what some might envision, natural selection is not a specific thing or single entity that is choosing or picking winners in the game of life. Rather, it is the natural outcome of the dynamic relationships that exist between an organism and all the various aspects of the organism's environment. This includes the relationship between the organism and other organisms within the same species, the relationship with other species within the organism's ecosystem, the relationship with abiotic factors such as climate and soil quality, and the chemical and physical constraints imposed by the structure of the universe we inhabit. All of these factors have a role in shaping the organism and determining how well it thrives. Complicating matters further, the organism, because it interacts with these factors, has a role in shaping and modifying many of them and, through various feedback loops, its own form.

Regardless of all the complexities that go into the process of natural selection, it has the ability to alter *populations* of organisms over time whenever the following three simple conditions are met. The first condition is that there is variation within the population: different members of the population look and behave differently. The second condition is that some of this variation is the result of the genetic information that an individual receives from its parents (as well as the occurrence of new genetic mutations). This does not mean that every aspect of a characteristic must be determined genetically, only that a genetic component plays a significant role in the development and expression of that characteristic. For example, one could have the genetic potential to reach a height of six feet, but one might not live in

20. For an easily readable overview of the EES, see Chiu, *Extended Evolutionary Synthesis.*

an environment in which one can obtain an adequate amount of nutrition to reach this height. This caveat aside, the third condition at work is that if there is, on average, a reproductive benefit of having a specific variation (tall height for instance), the genes associated with that phenotype will, on average, increase over time in the population.

Taken together, the fact that 1) individuals vary, 2) some of this variation is heritable, and 3) some of this heritable variation confers a benefit when it comes to survival and reproduction in the population's specific environment means that, over time, populations will undergo change and evolve naturally. Put simply, populations are shaped by their interactions with the environment over time. As the environmental interactions change—a new species invades an ecosystem, the climate gets warmer, etc.—the population will be reshaped accordingly, as different individuals within the population will have, on average, higher reproductive fitness (leave more offspring with their similar genetic makeup) than the individuals who might have done better under the original environmental conditions. This leads to a change in the genetic and phenotypic makeup of the *population* in subsequent generations.

One additional point about the nature of natural selection needs to be discussed here, and that is the nature of mutations. One aspect that fuels the development of species via natural selection is the introduction of new genetic material into populations. Many people describe the occurrence of new mutations as random and claim this undermines any belief that evolution via natural selection could be under the control of a providential God.

While this argument is addressed in detail in the next chapter, it is worth noting that mutations are not random. Randomness would imply they occur with no recognizable pattern. In fact, certain regions of the genome are known to have higher mutation rates than others, organisms have mechanisms to increase mutation rates in specific regions,[21] organisms can activate the movement of pieces of DNA called transposons in certain regions of the genome at specific times,[22] and organisms can direct DNA repair mechanisms to fix mutations in nonrandom manners that may facilitate evolutionary change.[23] While there is a probabilistic nature to the occurrence of mutations, mutation is *not* a random process. There is far more order and directness than previously thought.

Putting the question of the nature of mutations aside, very few disagree with the notion that natural selection operates in the natural world. In fact, even Young Earth Creationists agree that natural selection can shape and alter populations over time.[24] However, these critics contend that natural selection is limited in scope—that the changes mediated via natural selection are merely minor adjustments or tweaks to species and that large-scale evolutionary changes are beyond its scope. They would argue that natural selection can explain why lactose tolerance may have spread through a significant portion of the human population, but that it cannot explain such things as how a certain group of lobe-finned fish, over time, gave rise to well-adapted tetrapod forms. In fact, they would argue that no natural processes can explain this, neither natural selection nor the other mechanisms put forth by the EES.

In contrast, the naturalistic evolution position maintains that every evolutionary transition during natural history, all of the changes associated with universal common descent, can be explained by natural processes. As discussed earlier, demonstrating

21. R.S. Galhardo, P.J. Hastings, and S.M. Rosenberg, "Mutation as a Stress Response and the Regulation of Evolvability," *Critical Reviews in Biochemistry Molecular Biology* 42, no. 5 (2007): 399–435, https://doi.org/10.1080/10409230701648502.

22. N.H. Kim et al., "Real-time Transposable Element Activity in Individual Live Cells," *Proceedings of the National Academy of Sciences of the United States of America* 113, no. 26 (2016): 7278–7283, https://doi.org/10.1073/pnas.1601833113.

23. J.G. Monroe et al., "Mutation Bias Reflects Natural Selection in *Arabidopsis thaliana*," *Nature* 602 (2022): 101–105, https://doi.org/10.1038/s41586-021-04269-6.

24. For an example of this, see "Natural Selection," Answers in Genesis website, https://answersingenesis.org/natural-selection/.

this is not an easy task. Given that it is largely a historical science, making the case for naturalistic evolution is analogous to the efforts of a lawyer attempting to assemble enough circumstantial evidence to convict a murder suspect. Despite this difficulty, scientists have amassed a good deal of circumstantial evidence that is consistent with and supportive of natural selection and other natural processes driving evolution.

To illustrate the type of evidence that has been amassed, it is instructive to look at the transition from lobe-finned fish to land-dwelling tetrapods, a transition that is believed to have occurred roughly 360 million years ago (MYA). If one examines fossil beds dated to around 400 MYA, one finds that they are devoid of land-dwelling tetrapods. If one examines fossil beds dated to around 330 MYA, one finds a number of animals with four limbs that appear quite well adapted to living on land. According to evolutionary theory, natural processes should have driven this transition through a series of intermediate-looking forms that track to changes in the environment during this period. This prediction is actually borne out in the fossils.

If one examines rocks that are between 330 and 400 million years old, one finds a number of intermediate forms, odd-looking organisms that aren't quite land-dwelling tetrapods but seem to have some adaptations for life on land.[25] More than ten different intermediate forms have been found in fossil beds of this age, many in the last two decades. The heterogeneity of these forms makes it difficult to trace a nice linear transition within these intermediate fossils from aquatic to land-dwelling tetrapod. Rather, what these forms seem to indicate is that a good deal of evolutionary experimentation was taking place at this time with

25. The following is a good accessible overview of the tetrapod fossil record: Jennifer A. Clack, "The Fish–Tetrapod Transition: New Fossils and Interpretations," *Evolution: Education and Outreach* 2 (2009): 213–223, https://doi.org/10.1007/s12052-009-0119-2.

respect to producing land-dwelling tetrapod forms.[26] Why might this be the case?

It turns out that the environmental conditions of the time—lots of swamp-like areas with shallow streams and riverbeds—would have provided the exact environmental niche such that intermediate forms that developed limbs, even if they were not particularly adept at supporting one's body weight outside of water, would have had an advantage. Intermediate forms, even if they were not great at moving on land, would still have gained a benefit if they could access the nutrients available in very shallow water and on the land's edge. Even if they could only do so haltingly, by awkwardly moving on land for brief spurts, these new forms would have faced no well-adapted land-dwelling tetrapod competitors at the time.

Of course, as time went on and further evolutionary development occurred, forms better adapted to land would emerge from this bush-like evolutionary experimentation, and the initial odd-shaped tetrapod forms would go extinct. In fact, this is exactly what the fossil record depicts. The early tetrapod fossil form *Tiktaalik* (375 MYA) looks like a mix between a fish and a crocodile and may have only been able to lift its head out of water to eat on the banks of streams and rivers. Likewise, the tetrapod fossil form *Ichthyostega* (365 MYA) most likely moved very inefficiently, undulating like a seal across the landscape. Eventually these forms were replaced by more advanced tetrapods, ones that could fully support their body weight outside of water and locomoted in a manner more akin to modern amphibians.

Of course, just finding these fossils doesn't prove naturalistic evolution, but finding these intermediate fossils in this particular temporal sequence, less adapted forms emerging before more adapted ones, is consistent with natural processes honing the

26. J.A. Long and M.S. Gordon, "The Greatest Step in Vertebrate History: A Paleobiological Review of the Fish-Tetrapod Transition," *Physiological and Biochemical Zoology* 77, no. 5 (2004): 700–719, https://doi.org/10.1086/425183.

tetrapod form over time. In addition, recent laboratory work has supported the idea that selection could work on genetic and environmental modifications that have been shown by researchers to alter the limbs of aquatic species. For example, experiments have been done in zebrafish that show a *single* mutation in a developmental gene could give rise to the emergence of new bones and musculature in the limb, generating a limb that is remarkably similar to a primitive tetrapod limb.[27] In addition, lab experiments have shown that an aquatic ray-finned fish called *Polypterus* can develop a modified shoulder joint and propel itself on land simply by being raised in a terrestrial environment.[28]

The ability to create an intermediate form in the lab gives a good deal of support to naturalistic evolution, particularly the notion that the specific genetic variation necessary for this transition would likely have been available for selection to act upon. However, even this does not *prove* that naturalistic evolution is correct. The reality is that there is likely no conceivable evidence that will do this given the historical nature of the questions at hand. One cannot rerun the evolutionary tape with the identical environmental conditions and observe exactly what occurred. If one is looking for certainty, one will not find it in evolutionary science.

But, as discussed previously, certainty cannot be the goal. Rather, one should be asking, given the evidence we have now at our disposal, what scenario or hypothesis best makes sense of it. As more and more intermediate forms are unearthed and as we learn more about the genetic and developmental underpinnings that shape organismal form, the data turns out to be largely consistent with the expectations of an evolutionary transition driven by natural processes.

27. This is consistent with the EES idea of Evo-Devo described above (M.B. Hawkins et al., "Latent Developmental Potential to Form Limb-like Skeletal Structures in Zebrafish," *Cell* 184, no. 4 [2021]: 899–911).

28. This is consistent with the EES idea of phenotypic plasticity described above (E.M. Standon et al., "Developmental Plasticity and the Origin of Tetrapods," *Nature* 513 [2014]: 54–58).

NATURALISTIC EVOLUTION AND CATHOLIC CREATION

The prospect that evolutionary common descent could be explained via natural processes tends to encourage people to jump to unjustified theological or philosophical conclusions. For example, many evolutionary biologists argue that if we can provide, in theory, a completely natural account of evolutionary history, this would remove the need for God from the picture. Moreover, religious critics of naturalistic evolution make this same assumption, an assumption that fuels their animosity toward evolution.

The proposed connection between naturalistic evolution and atheism is illustrated in the following quote from the evolutionary biologist Douglas Futuyma: “If the world and its creatures developed purely by material, physical forces, it could not have been designed and has no purpose or goal. . . . Some shrink from the conclusion that the human species was not designed, has no purpose, and is the product of mere mechanical mechanisms—but this seems to be the message of evolution.”[29]

But is this conclusion that Futuyma reaches a scientific conclusion or a philosophical one? Does the scientific data demonstrate his conclusion, or does his conclusion flow from certain philosophical presuppositions that he fails to acknowledge? If one examines his reasoning closely, one finds that he makes the erroneous assumption that if something can be explained via material processes, this by necessity removes God from the picture. Yet, as discussed in chapter 3, God is not competing with natural causes in this manner. Rather, God is the primary cause of all that exists, and physical forces and processes only have efficacy because God has endowed them with existence and efficacy. As a result, explanations that rely on “material, physical forces”

29. Douglas Futuyma, *Science on Trial: The Case for Evolution* (New York: Pantheon Books, 1982), 12–13.

are entirely consistent with God. There may be limits to what material and physical forces can do, but if they are found to be capable of producing new species, they do so only because God 1) has intended them to do so and 2) has endowed them with the powers to produce new species.

As a result, if one were to demonstrate that "creatures developed purely by material, physical forces," would this situation necessarily imply atheism or purposelessness? Of course not. It only leads one to such conclusions if one is already committed *a priori* to them. It is quite possible that God could have created a world in which natural forces alone could not only produce living organisms of all kinds but produce exactly the kinds of living organisms that he desired. Such a world would have been designed from the very beginning for his purposes. Given this alternative possibility, Futuyma's conclusion does not follow by necessity from the science.

Unfortunately, the philosophical overreach demonstrated by Futuyma and others like him goes even deeper. In assuming that everything in the universe, including humankind, can be entirely explained by purely material explanations, they, whether consciously or not, assume that modern empirical science can answer all meaningful questions. Futuyma assumes that man is nothing more than a collection of molecules, that he is *not* a unity of body and soul, and that our rational abilities do not transcend our material being. He never justifies these presuppositions, which go well beyond what the science can demonstrate. His entire quote is a philosophical argument that is based upon materialistic philosophical assumptions that he never bothers to justify. What it would be more intellectually honest for him to say is the following:

> I believe that the world and its creatures developed purely by material, physical forces. This is something that I can't verify scientifically, as I am taking it on faith that humans are merely

> material beings. If I am correct and the world consists only of material substances that developed via material forces, this would be consistent with what one would expect of a world devoid of a Creator or purpose or plan. It doesn't prove there is no Creator, though, because it is still possible that God could have created a world that could be explained by nothing but physical forces. However, I believe that this is unlikely because I'm not very keen on the idea of God.

This much more modest proposition is really all atheistic evolutionists can claim, making the supposed link between atheism and naturalistic evolution appear much more tenuous. They also have no way of explaining why this particular highly ordered universe, the rare universe that is able to support evolutionary processes, should come to exist in the first place. It should be clear, hopefully, that the mere fact that a scientific account of evolution could explain the development of the material aspects of organisms (including man) does not undermine the faith. The second creation account in Genesis depicts God forming man from the dust of the ground. On Ash Wednesday, we are reminded of this through the incantation of the scriptural phrase "Remember that you are dust, and unto dust you shall return." The Church reminds us that we are physical beings and therefore are part of the physical world. Thus, it is not antithetical to our faith if the science of evolution is able to explain the development of our physical structure, the proverbial dust that God used to bring us into existence. Such an explanation does not exhaust what it means to be human because we are more than dust; we are more than physical beings. We have, as described in the Genesis account, the "breath of life" that animates us. We possess the life-giving spirit of God. We are a unity of body and soul, and any scientific evolutionary description, if it is to remain a purely scientific account, cannot adequately address the immaterial aspect of our being.

6

The Order of Evolution

Reconciling Chance and Purpose

The general forms that life can develop and adapt are not haphazard but follow definite chemical, genetic, and environmental pathways. . . . In other words, there is a deeper structure that makes the adventure of biological life not utterly random but orderly, somewhat like jazz music in which basic tunes . . . are recognizable when played, but are always played with innovation and creativity.[1]
—Christopher Baglow

Each of us is the result of a long chain of what we would popularly refer to as chance events. Each of us came into being as the chance encounter of one specific ovum with one specific sperm. Furthermore, that specific sperm and specific ovum came to carry our own unique DNA mix via the result of many chance events that occurred during the process of sperm or ovum formation. In addition, we are also the result of the chance encounter that brought our parents together, who themselves are the result of the chance encounters that brought their parents together, and so on. Any cursory glance at our individual histories reveals a staggering number of chance events upon which our existence is predicated. Yet, this seems to set up an apparent dichotomy because as Catholics we believe that we have been known by God for all eternity, that before he formed us in the womb, he knew us.

Such a glance at our own personal history should make it abundantly clear that the issue of reconciling chance events with divine providence is not something that only surfaces in relation to evolution. The issue is merely magnified with evolution because

1. Christopher Baglow, *Creation: A Catholic's Guide to God and the Universe* (Notre Dame, IN: Ave Maria, 2021), 50–51.

evolutionary data reveals that chance events have influenced our emergence as a species as a whole. Had not certain chance events occurred during evolutionary history, it seems that the species known as *Homo sapiens* might never have come into being. Given this, are we merely the chance by-product of a blind evolutionary process that did not have us in mind, or are we the intended creatures of a loving Creator who designed and fashioned us with care? Which one is it?

Darwin himself struggled with this apparent dichotomy, writing the following in one of his letters to the American botanist Asa Gray: "I am conscious that I am in an utterly hopeless muddle. I cannot think that the world, as we see it, is the result of chance; and yet I cannot look at each separate thing as the result of Design."[2]

Darwin only seemed to see two options: either every contrivance in the natural world, from the beak contours of a finch to the coloration of a fish, was the result of direct divine intervention (as the natural theologians of Darwin's time concluded), or the ordered creatures we find around us are the result of mere happenstance. Is there a middle ground that allows one to accept the chanciness of an inherently messy evolutionary process but still retain the belief in the providential guidance of a divine Creator? From a Catholic perspective, there must be, because without a way of integrating chance and providence, not only would evolution fall outside God's guidance, but our very existence, our chance formation in the womb, would fall outside of it as well. There is a need then for Catholics to reconcile chance and providence irrespective of the truth of evolution. So it should not be surprising that this is a topic the Church addressed well before the advent of modern evolutionary theory.

2. Charles Darwin, "Letter to Asa Gray," November 26, 1860, Darwin Correspondence Project website, https://www.darwinproject.ac.uk/letter/DCP-LETT-2998.xml.

CHANCE VS. RANDOMNESS IN EVOLUTION

Before discussing how chance and providence can be integrated, it is useful to clarify what is meant by chance in evolution and to separate this from claims regarding randomness. Notice that in the above quote, Darwin used the word chance rather than randomness. In fact, the words random or randomness do not appear anywhere in Darwin's *Origin of Species*. This is a key point because although the terms randomness and chance are often used interchangeably in discussions regarding evolution, they represent two distinct concepts.

Randomness, from a scientific perspective, refers to a series of events that do not follow any type of pattern. Given a truly random list of digits, one is unable to predict or ascertain any pattern. For example, knowing that the first four digits of a truly random series of numbers are 3, 8, 2, and 9 would give you no information regarding the likelihood of whether the next digit in the series is a 4. Evolution is not random in this sense, as there are levels of patterns and probabilities such that some level of prediction is indeed possible.

First, mutations do not occur with equal likelihood throughout an organism's genome. In fact, there are regions known as mutational hotspots in which mutations are much more likely to occur. In addition, there are environmental stresses that increase the likelihood of mutations occurring in specific regions of an organism's genome. Recent research has even shown that there are more new mutations that occur in the gene associated with sickle cell anemia in Africans who live in malaria-infested regions than occur in Caucasians.[3] Given that heterozygous mutations in this gene can give a person resistance to malaria, mutations in this gene are clearly not occurring randomly with respect to fitness.

3. D. Melamed et al., "De Novo Mutation Rates at the Single-Mutation Resolution in a Human *HBB* Gene Region Associated with Adaptation and Genetic Disease," *Genome Research* 32, no. 3 (2002): 488–498.

Rather, they occur with a higher frequency in the individuals in whom the mutations would be more likely to be evolutionarily adaptive, Africans living in malaria-infested areas.

Likewise, larger DNA alterations that are driven by pieces of DNA called retrotransposons that can move around in the genome also are not random.[4] These retrotransposons seem to be mobilized preferentially at different times and in different regions of the DNA. There seems to be a bias for retrotransposon activity during certain stages of development and a bias for insertion upstream of gene sequences in DNA regions that control the activity of the gene.[5] Again, the process seems far from random.

The other aspect of mutational change in an organism's DNA that is not random is the fact that once mutations or DNA damage occurs, organisms mount robust yet varied responses to the damage. These include repairing and minimizing the effect of the mutation, creating additional mutations during the repair process, or allowing a specific mutation to persist. Mutations that occur in certain locations or in specific genes are more likely to be repaired than others. The reality is that mutations occur in a non-random fashion and are shaped by very fine-tuned organismal processes, processes that have evolved over time in order to promote the survival of the organism.

While the mutations that shape the evolutionary process are not random, there are many events within the evolutionary process that can be attributed to chance. While randomness, as previously described, is associated with events that lack any predictable pattern, chance events, within an evolutionary context, correspond to the intersection of two causally unrelated events. This does not imply that the two interacting events do not follow regular patterns or are totally unpredictable on their own

4. G. Bourque et al., "Ten Things You Should Know about Transposable Elements," *Genome Biology* 19, no. 199 (2018): https://doi.org/10.1186/s13059-018-1577-z.

5. Y. Low et al., "Transposable Element Dynamics and Regulation during Zygotic Genome Activation in Mammalian Embryos and Embryonic Stem Cell Model Systems," *Stem Cells International* (October 15, 2021), https://doi.org/10.1155/2021/1624669.

terms. It simply refers to the fact that their causal histories are unrelated to each other, despite the fact that they interact.

An evolutionary example should help illustrate this point: the chance occurrence of DNA damage via a chemical mutagen in a frog. In this example, the chemical mutagen has its own causal history regarding how it was produced in the environment, how it came to be located in a specific pond, etc. The frog also has its own causal history regarding its formation as a zygote, its life history, its current search for food near the edge of the pond, etc. They both have their own independent causal histories that can explain how they came into contact in the same pond on a particular day. Because of their independent causal histories, their encounter would be considered a chance event. The mutagen was not looking for the frog, nor was it attracted to the frog and vice versa. Both had separate, independent causal histories that caused them to be in that same location on a specific day. The fact that these two, the chemical mutagen and the frog with its specific mutagen response pathways, intersected in such a way that DNA damage eventually occurred would be considered a chance event.

In such an event, the frog does not get to pick which region of the DNA will encounter the mutagen (although some regions will be actively protected from damage while other regions will be more susceptible to mutagens). It becomes a matter of probability then regarding which DNA region any specific mutagens might encounter. Such probabilistic chance encounters occur repeatedly during the evolutionary process.

In addition to the element of chance involved in DNA alterations, evolution is also influenced by chance environmental perturbations. Many changes in ecosystems that affect the survival of organisms are triggered by chance events such as volcanic eruptions, meteor collisions, or the breaking of a natural dam. The organisms that live near a large-scale volcanic eruption have their specific causal history of how they came to exist in that specific locale, but by chance, they must deal with the environmental consequences of the eruption, which has its own separate causal history. Such chance environmental disturbances can have wide-ranging effects on the organisms in their vicinity. As a result, just like DNA changes, such events can have a significant effect on the direction of evolution.

PRIMARY AND SECONDARY CAUSALITY: CHANCE AND PURPOSE

It seems clear that the process of evolution is dependent upon and influenced by chance events. How then can it be reconciled with God's providential design? From a scientific perspective, it might seem that chance and design are mutually exclusive. Either markings on a piece of ochre from a prehistoric archeological dig were deliberately made by hominins, or the markings were merely chance scratches made when the ochre was hit by a falling rock. Likewise, holes found in a prehistoric animal bone either were deliberately made in the bone to fashion a musical instrument or were the result of chance punctures as animals gnawed on the bones for sustenance.

Chance and design seem to be mutually exclusive explanation; therefore, there is a perception that naturalistic theories of evolution have dealt a death blow to the possibility of design in the natural world. Interestingly, though, in the *Origin of Species*, Darwin seems to leave open the possibility of divine providence acting within creation. However, his position on exactly how this might occur is somewhat unclear. Both in the quotation from William Whewell that he placed opposite the title page and in his famous line about a Creator at the end of the book,[6] Darwin seemed to propose that God had ordered a world according to fixed laws. Under this interpretation, God is seen as a watchmaker

6. "There is grandeur in this view of life, with its several powers having been originally breathed by the Creator into a few forms or into one; and that, whilst this planet has gone cycling on according to the fixed law of gravity, from so simple a beginning endless forms most beautiful have evolved" (Charles Darwin, *The Origin of Species* [New York: Mentor, 1958], 459).

who set the universe in motion, providing it with the "laws impressed upon matter by the Creator."[7] While this deistic interpretation allows some sort of design into the equation, the role of divine providence and God's sovereignty over creation is significantly diminished. While God may have designed our specific universe at the very onset, what it actually produces via an evolutionary process would be left to happenstance. For example, such a deistic God would seem to have no direct hand in the chance events that led to a relatively hairless bipedal hominid appearing in Africa some 200,000 to 300,000 years ago. In this distinctly non-Catholic reading, the universe may be the result of design, but what it has produced is the result of chance.

How then does one reconcile a Catholic understanding of God, who is sovereign over all his creation, with a chance process that appears to have a "mind" of its own? This is where a proper understanding of the Church's view of God as Creator is helpful. As discussed in chapter 3, the creative power of God is such that he is the ultimate or primary cause of all that exists. It is simply not proper to think in either/or terms: either God is the cause of evolution or natural processes act as the cause. Either God is the cause of this species changing into another species or natural processes are the cause. Such thinking reduces God to one competing cause among many others rather than viewing him as the first and ultimate cause of all that exists. As the philosopher William Carroll explained, "God's causality is so different from the causality of creatures that there is no competition between the two, that is, we do not need to limit, as it were, God's causality to make room for the causality of creatures. God causes creatures to be causes."[8]

In this sense, while I am the cause of this book as its author, I am only able to produce this chapter because God sustains my

7. Darwin, *Origin of Species*, 458.

8. William Carroll, "Stephen Hawking's Creation Confusion," *Public Discourse*, September 8, 2010, https://www.thepublicdiscourse.com/2010/09/1571/.

existence *at each and every moment.* Not only that, he sustains my existence as the *specific type* of creature, a rational human, that has the conceptual ability to produce a book. The same primary/secondary causality holds true for chance events. A chemical mutagen that binds to the DNA of a specific sperm cell within a frog is not competing with God as a cause, but rather is participating in the primary causality of God. God is the ultimate cause of the mutagen's existence as the type of entity that can affect the DNA in a specific manner based upon the underlying chemistry. Likewise, the frog is sustained in existence by God, which allows the frog to do what frogs naturally do—eat, avoid predators, fertilize eggs, etc. When these things operate according to their own natural properties, they may, by chance, encounter each other such that a mutation occurs. Each is operating in a regular manner based upon the type of entity that it is, either as a chemical or a frog. Because God, as the primary cause, allows such entities to act according to their natures, such chance events are not opposed to God's designs or providence. In fact, such chance events only occur because God, as primary cause, allows chemicals and frogs, as secondary causes, to operate in the manner that they typically do.

This distinction between primary and secondary causality has a long history within the Catholic intellectual tradition and was summed up succinctly by Thomas Aquinas over five hundred years ago: "The effect of divine providence is not only that things should happen somehow, but that they should happen either by necessity or by contingency. Therefore, whatsoever divine providence ordains to happen infallibly and of necessity happens infallibly and of necessity; and that happens from contingency, which the divine providence conceives to happen from contingency."[9]

What Aquinas is claiming is that chance events (he uses the term contingency) occur only because God as the first cause of all that exists allows them to occur. They cannot occur independently

9. Thomas Aquinas, *Summa theologiae* 1.22.4 ad. 1.

of God because the entities involved in the chance encounter only exist and operate because God sustains them in their ability to do so. Therefore, anything they do, including any chance encounters in which they are involved, cannot occur outside of his providential plan. Within the Catholic tradition, chance events in evolution or chance encounters that occur in everyday life are not somehow left outside God's care. In fact, everything God has created, from gravity to electrons to turtles, has inherent abilities to act as a causative agent, abilities that owe their origin and continued sustenance to God as a primary cause. As the mathematician James Bradley explains, "This view does not deny that God is in control—it simply says that God exercises control not by micromanaging (occasionalism) but in other ways including establishing properties, capabilities, and limitations for creatures."[10]

Interestingly, such a notion was not foreign to Darwin, as he and the American biologist Asa Gray frequently discussed these matters in the letters they exchanged. Gray once wrote that "God himself is the very last, irreducible causal factor and, hence, the source of all evolutionary change."[11]

Gray, though, seemed to go even further than this. Rather than merely seeing God as the primary cause of evolutionary events, Gray also seemed to indicate in his writing that God intervened to lead the chance variations down specific pathways. Darwin for his part clearly viewed this notion as problematic. In one letter, he responded to Gray by stating, "I cannot look at each separate thing as the result of Design.—To take a crucial example, you lead me to infer that you believe 'that variation has been led along certain beneficial lines.' I cannot believe this and I think you would have to believe that the tail of the Fan-tail

10. James Bradley, "Random Numbers and God's Nature," in *Abraham's Dice: Chance and Providence in the Monotheistic Traditions*, ed. Karl W. Giberson (Oxford: Oxford University Press, 2016), 76.

11. Asa Gray, quoted in George Webb, *The Evolution Controversy in America* (Lexington: University Press of Kentucky, 2002), 19.

pigeon was led to vary in the number and direction of its feathers in order to gratify the caprice of a few men."[12]

But while in Darwin's view God may not have been directly intervening to create specific mutations, Darwin did see evolution as being driven or shaped by the laws of nature, laws that owed their existence to the Creator. In this way, his views in the *Origin* were, in a certain respect, compatible with Bradley's view of God exercising guidance and control through the manner in which the world, its physical forces, and its myriad created causative agents operate, even if Darwin never explicitly made this connection.

For Catholics, though, it is important to not interpret Bradley's view as in any way approximating a deistic view of God, one in which God endows his creation with order and allows it to run its course unchecked. Rather, Bradley's view is the complete opposite. It is one in which God is deeply present to his creation at each and every moment. It is one in which God is intimately united with his creation because everything that occurs, occurs precisely because God allows and sustains its occurrence. The theologian Matthew Ramage explains the implications of this relationship:

> The fact that God achieves his end for the universe by working through the natures of his creatures not only manifests his wisdom and power; it also bespeaks his goodness. That is, God's will to rely so deeply on creaturely causation within evolution blesses us creatures with the immense dignity of participating as his coworkers in the governance of creation. As we have always seen in such practices as intercessory prayer to the saints, the Catholic impulse is to magnify the creature's participation in God's action rather than be ashamed of it. . . . In a nutshell, it is more perfect and fitting for God to make

12. Charles Darwin, "Letter to Asa Gray," November 26, 1860, Darwin Correspondence Project website, https://www.darwinproject.ac.uk/letter/DCP-LETT-2998.xml.

> creatures authentic causes of evolutionary change in virtue of their own natures [created and sustained by God] than for him to remain the sole cause of new life in the universe.[13]

God is the sole creator of all that exists, in that he brings and sustains all things in existence. However, as created entities change during the evolutionary process to produce new species, these changes can be driven by created causes like chemicals, frogs, sunlight, and bacteria. They do this precisely because they act in accord with their God-given natures. As a result, out of the chance encounters of these orderly created causes, God, in concert with his creation, can bring forth new lifeforms.

CHANCE VS. ORDER IN EVOLUTION

Throughout the *Origin*, it is particularly striking how many times Darwin stresses the order and law-like properties of the evolutionary process. One of the most telling passages is found near the end of the book, in which he sums up the outcome of natural selection by stating, "These elaborately constructed forms, so different from each other, and dependent on each other in so complex a manner, have all been produced by laws acting around us."[14]

It is important to note that he did not state that these forms "were cobbled together by chance and happenstance." Rather, he goes so far as to state that given the action of these laws around us, "the most exalted object which we are capable of conceiving, namely, the production of the higher animals, directly follows."[15] While he did not deny the role of chance variations, Darwin did not attribute the evolutionary process to mere happenstance. One quote from the *Origin* is quite striking in

13. Matthew Ramage, *From the Dust of the Earth: Benedict XVI, the Bible, and the Theory of Evolution* (Washington, DC: The Catholic University of America Press, 2022), 80–81.

14. Darwin, *Origin of Species*, 459.

15. Darwin, 459.

this respect. In discussing the chance variations that occurred in organisms, Darwin stated, "I have hitherto sometimes spoken as if the variations—so common and multiform in organic beings under domestication, and in a lesser degree in those in a state of nature—had been due to chance. This, of course, is a wholly incorrect expression, but it serves to acknowledge plainly our ignorance of the cause of each particular variation."[16] Darwin, for his part—again, while recognizing the chanciness of evolution—seemed more interested in uncovering the underlying laws and order and patterns of evolution.[17]

This perspective is often lost in current discussions regarding evolution, largely because the modern popularizers of evolutionary thought tend to exalt chance above all else. Yet, in the case of evolution, it is the order that exists in nature that is primary, and it is the chance aspects of the process that are secondary. In fact, the chance aspects of evolution, by and large, operate in such a manner as to uncover the order that is already hardwired into creation.

To understand the relationship between order and chance in evolution, the game of roulette serves as a useful analogy. When one spins the roulette wheel, the ball is equally as likely to fall into the red three slot as it is to fall into the black four slot. The odds are 1 in 38 for any specific number/color combination occurring. However, for the game to work in this manner, the roulette wheel must be constructed in a very specific way with a high degree of order. Without this underlying order and structure in the roulette wheel, the game would be totally different. If the red three were twice the size of the black four, then the game would be altered based upon this new underlying order in the altered roulette wheel. It is only the order of the roulette wheel that allows the

16. Darwin, 133.

17. Darwin uses chance in a variety of different contexts in the *Origin*, so there is considerable debate regarding how he understood chance. What seems clear, though, is that he thought that there were undiscovered laws and patterns that influenced and could explain many aspects of evolution.

game to function properly as the game of chance we recognize. Furthermore, if the red three slot was unstable such that its size was continually changing (maybe the wheel was poorly constructed such that the slots continually moved or tilted), one couldn't even have a true game of chance at all.[18]

In a similar manner, for the chance aspects of evolution to assist in the evolution of organisms, a high degree of order must already exist in the underlying physics and chemistry of our universe. In the case of evolution, without the law and order evident in the chemical and physical regularities of the world around us, evolution could never get off the ground. Without the foundational order that has been revealed by modern physics, there would be no stable atoms or molecules, the raw material that is absolutely essential for evolution to work at all. And the order in carbon, oxygen, and nitrogen atoms is what allows for the amino acids, the building blocks of proteins that are composed largely of these atoms, to be produced abiotically (spontaneously outside of living organisms), as they have been found throughout the universe in meteors, asteroids, and in gas clouds surrounding star-forming regions.[19]

As the physicist Stephen Barr eloquently put it, "The order which we see in nature at one level has its roots in a more mathematically perfect order that exists at a deeper level."[20] Each layer is dependent upon the layer beneath it. Without stable atoms, one could not have stable molecules like amino acids. Without stable molecules, there could be no stable organisms, which are

18. In addition, even though roulette is a game of chance, the casino can use this for a definite purpose—to make large sums of money. Chance processes, the outcome of the ball on the roulette wheel, can be used for a purpose. A similar situation occurs with many computer algorithms that use an element of chance to perform a specific task. There is no inherent contradiction between chance events and a purposeful process.

19. Susana Iglesias-Groth, "A Search for Tryptophan in the Gas of the IC 348 Star Cluster of the Perseus Molecular Cloud," *Monthly Notices of the Royal Astronomical Society* 523, no. 2 (2023): 2876–2886, https://doi.org/10.1093/mnras/stad1535.

20. Stephen Barr, *Modern Physics and Ancient Faith* (South Bend, IN: University of Notre Dame Press, 2003), 87.

the objects of natural selection. Without these, there is no such thing as biological evolution.

Change this order slightly, and everything falls apart. If the strong nuclear force were weakened slightly, it would lead to a universe composed only of hydrogen atoms. Such a universe would be completely devoid of complex molecules and organisms. This would be a universe that could not support any type of biological evolutionary process. The astonishing reality is that we happen to occupy a universe in which the physics is ordered in such a manner that there are roughly eighty different stable elements that can combine based on the underlying chemistry to produce the organic molecules necessary for the production of living organisms. This degree of order at the level of the physics is not trivial, as it involves many specific and highly particular values, such as the size of the proton, the magnitude of the weak and strong nuclear forces, etc. Furthermore, this spectacular underlying order did not evolve; it simply *is* the structure of our universe. As such, it represents a gift to evolution, a gift that not only allows evolution to occur but, as Darwin seemed to intuit, also shapes its outcomes.

One of the most interesting examples of how the underlying chemical and physical order shapes evolution can be seen at the level of proteins. Proteins are one of the key macro-molecules found in living cells, and they perform many functions that are essential for life. Biological proteins are strings of amino acids connected in long chains, and there are roughly twenty different types of amino acids that can be strung together to make biological proteins. Just like the letters of the alphabet can be arranged to produce words and sentences, amino acids can be arranged to produce proteins. Since typical proteins are usually hundreds of amino acids long, the number of possible protein sequences is virtually infinite, given that there are twenty possible amino acids one could have at each position within the protein. Even for a protein that is only one hundred amino acids long, this would mean that there are 20^{100} possible protein sequences. Based on this seemingly infinite variety of possible amino acid sequences, one might think that the specific three-dimensional protein structures that evolution has stumbled upon are simply the result of chance. However, this does not seem to be the case. It turns out that, based upon the underlying chemistry, the possible ways in which chains of amino acids (proteins) can fold are limited.

Despite the existence of a nearly infinite array of possible amino acid sequences, proteins predominately fold into *two* distinct secondary structures: beta sheets and alpha helices. They do this because these are the two main stable secondary structures that strings of amino acids can form. For chemical and physical reasons, chains of amino acids in the cellular environment only tend to form stable structures if they adopt one or a combination of these two forms.[21] These forms are not built by the evolutionary process but are rather dictated by physical and chemical constraints. If evolution were to be run again, beta sheets and alpha helices would certainly form any time amino acids are strung together in a chain in an aqueous environment because that is what the chemistry dictates.[22]

Furthermore, the chemistry and physics dictate that beta sheets and alpha helices can only interact with each other in a limited number of ways. As a result, proteins fold into a rather limited array of three-dimensional structures. As the biologist Michael Denton has explained, "a number of organization rules, 'laws of form,' which govern the local interactions between the main structural submotifs [the alpha helices and beta sheets for example] have been identified, and these restrict the spatial

21. This is largely due to hydrophobic/hydrophilic interactions along the backbone of the protein chain that stabilize the structure of the protein in the cellular environment.

22. This also makes it much more likely to get a properly folded protein from a random string of amino acids. Instead of having to stumble on a proverbial needle in a haystack sequence (1 in 10100), recent research has shown that proteins that fold into alpha helices appear to emerge from random sequences at a much higher rate (1 in 102) (V. Tretyachenko et al., "Random Protein Sequences Can Form Defined Secondary Structures and Are Well-Tolerated *In Vivo*," *Scientific Reports* 7, no. 15449 [2017], https://doi.org/10.1038/s41598-017-15635-8).

arrangement of amino acid polymers to a tiny set of about 1000 allowable higher-order architectures."[23] Rerun evolution again using the same chemistry and physics, and similar folds would likely reappear because the structure of the folds is dictated by the chemistry and physics and not by happenstance.

Here is a concrete manner by which chance variations in protein sequences could be led down certain paths, not by the constant meddling hand of a Creator but via the inherent order embedded within creation. The order here drives the formation of certain stable protein folds, while the chance variations that occur in the amino acid sequence of the protein are secondary. These chance variations represent the engine used to uncover the protein fold possibilities dictated by the underlying order, possibilities that are in some sense written into the structure of the universe. This is a structure and order that, as the *Catechism* highlights, points toward the divine: "By the very nature of creation, material being is endowed with its own stability, truth, and excellence, its own order and laws. Each of the various creatures, willed in its own being, reflects in its own way a ray of God's infinite wisdom and goodness."[24]

Interestingly, the degree to which this order directs and guides the evolutionary process is not merely limited to the molecular level. In fact, there is evidence that evolution is constrained and channeled toward certain biological forms on the macro level as well.

23. Michael Denton, Peter Dearden, and Stephen Sowerby, "Physical Law Not Natural Selection as the Major Determinant of Biological Complexity in the Subcellular Realm: New Support for the Pre-Darwinian Conception of Evolution by Natural Law," *BioSystems* 71 (2003): 297303, at 299.

24. *Catechism of the Catholic Church* 339.

CONVERGENCE IN THE EVOLUTION OF ORGANISMAL FORMS

Given the myriad chance environmental perturbations and DNA insults that have influenced the roughly four billion years of evolutionary history, some, like the late paleontologist Stephen J. Gould, have argued that the evolutionary process is highly contingent. In his book *Wonderful Life: The Burgess Shale and the Nature of History*, he argues that, given the historical contingencies that effect evolution, if one were to rerun its four-billion-year tape a totally different set of organisms would emerge in this new iteration. There are others, though, who dispute this view and argue for a more convergent view, one in which similar organisms would emerge if one could rerun the evolutionary tape. The most vocal proponent of this view is the Cambridge paleontologist Simon Conway Morris, who argued in his book *Life's Solution: Inevitable Humans in a Lonely Universe* that there is an underlying order that in some sense directs or channels evolution. Conway Morris argues that regardless of the exact pattern of chance mutations and environmental changes, the underlying order will influence the direction of evolution such that it will converge upon certain similar, highly robust forms. The chance events allow the evolutionary process to search for and produce "endless forms most beautiful,"[25] but the forms it can produce are to some extent dictated by the underlying order. In a sense, the chance search uncovers what is already there. According to Conway Morris, "[Evolution's] central paradox revolves around the fact that despite its fecundity and baroque richness life is also strongly constrained. The net result is a genuine creation, almost unimaginably rich and

25. Darwin, *Origin of Species*, 459.

beautiful, but one with an underlying structure in which, given enough time, the inevitable must happen."[26]

While various arguments have been put forth regarding exactly how convergent or contingent the evolutionary process actually has been on our planet, it is important to recognize that the correct answer lies somewhere between the two extremes. Evolution is both contingent and convergent; the debate is really about which one is more predominant. Given our inability to rerun the four-billion-year process of evolutionary history in a laboratory, it might seem difficult to answer or address this question in any substantial manner. This is not quite the case though. In fact, by examining how the evolutionary process has played out on our planet, one can uncover a good deal of evidence supporting the notion that evolution is highly convergent (much like the convergent nature of protein structure discussed above).

The reality is that in the one iteration of evolution that has occurred on our planet, convergence in organismal form and behavior is ubiquitous. A convergent structure or behavior is defined as one that has evolved multiple times in separate independent lineages, and it turns out that there is hardly a structure or behavior that one can find in living organisms that is not convergent. For example, multicellularity has evolved multiple times, and so too has the complex camera-like eye that we possess. This eye structure has evolved at least seven different times, including in vertebrates, cephalopods, marine annelids, gastropods and even jellyfish.[27] Echolocation, powered flight, bipedal locomotion, the fusiform shape for swimming, and warm-bloodedness are just a few other examples of traits that have all evolved multiple times. Again, as Conway Morris points out, "The ubiquities of convergence are clear enough evidence that life has a peculiar

26. Simon Conway Morris, *Life's Solution: Inevitable Humans in a Lonely Universe* (Cambridge: Cambridge University Press, 2003), 20.

27. Conway Morris, 151–158.

propensity to 'navigate' to rather precise solutions in response to adaptive challenges."[28]

Even instances put forth as examples of radical contingency, such as the diversification of the mammals that occurred roughly sixty-five million years ago, may not be as clear cut as advocates of the contingent view claim. The diversification of the mammals occurred shortly after a mass extinction occurred at the end of the Cretaceous period—the K-T event. This extinction event is widely believed to have been triggered by an asteroid slamming into what is now known as the Yucatan peninsula. Before the K-T event, the dinosaurs were the predominant predators on the planet, occupying a wide range of ecological niches. In fact, the fossil layers of the upper Cretaceous right before the K-T event are filled with a diverse array of large dinosaurs. However, once one crosses into the Paleogene rocks right after the K-T event, the dinosaurs completely disappear. The mammal fossil record shows a distinctly different pattern, as the Cretaceous mammals, though diverse in nature, tend to be small in size and rather unimpressive. However, once one crosses into Paleogene rocks, the mammals radiate in both size and form.

The common "contingent" interpretation of this data is that without the chance asteroid impact, the dinosaurs would still reign supreme and mammalian evolution would have never taken off. As a result, evolution would have been pushed down an entirely different trajectory, one that would not have led to a bipedal mammalian hominin known as *Homo sapiens*. The standard contingent view is that we, as a species, owe our existence to this chance asteroid strike.

Yet there is an alternative explanation that, while not denying the role of chance in the evolutionary process, gives priority to the underlying physical and chemical order upon which evolution operates. From a convergent point of view, there is evidence that the mammal way of living is so robust given the physical

28. Conway Morris, 308.

principles of our world that eventually mammals would have displaced the dinosaurs as the dominant predators on the Earth. Under this interpretation, what is contingent is the *timing* of the rise of mammals, not the fact that they eventually evolved to prominence. In fact, the mammalian way of "making a living"—endothermy (the ability to maintain an elevated body temperature), giving birth to live young, and complex dentation—may have allowed for an adaptability that the dinosaurs simply lacked, and this, not chance, is the reason some mammals survived the K-T event while the dinosaurs did not.

For example, the fact that dinosaurs laid eggs made their offspring particularly vulnerable to environmental changes. (It also is one factor that allowed them to evolve to such enormous sizes.[29]) On the other hand, the fact that mammals give birth to live young and invest more resources in nourishing their offspring likely gave them a greater chance of surviving in an ever-changing environment (although this reproductive style limits body size). It is quite possible that the same factors that led to the rise of the dinosaurs, the large and specialized body types, eventually led to their inevitable undoing. Given the continual recurrence of massive environmental disasters in the fossil record, rerun the evolutionary tape and it is certain that these different animal groups would have had to deal with regular (on a geological timescale) massive environmental upheavals similar in scope to the K-T event. Certainly, the timing of such events or the exact environmental trigger would change in any rerunning of the tape, but a similar outcome would be reached: the eventual establishment of mammal-like creatures as the dominant predators on the planet because they have an advantage in adapting to large-scale environmental change.

While both the contingent and the convergent interpretations regarding the rise of the mammals are logically possible,

29. P. Sander et al., "Biology of the Sauropod Dinosaurs: The Evolution of Gigantism," *Biological Reviews* 86, no. 1 (2011): 117–155. https://doi.org/10.1111/j.1469-185X.2010.00137.

how does one distinguish between them given that rerunning the tape is not a possibility? One manner in which to get at this is to investigate whether the evolutionary process is convergent in respect to the mammalian form: Does it produce the characteristics associated with mammals multiple times in various independent lineages? Basically, one can ask, is the mammalian form convergent? If, based upon the underlying order and structure of our world, the mammalian form is one of those "rather precise solutions," then one might expect mammalian traits to evolve multiple times independently. As it turns out, this type of convergent evolution is exactly what one finds.

One structure that makes mammals unique is their possession of a specific set of inner ear bones, the malleus, incus, and stapes, which are believed to have evolved from portions of the jawbone of mammalian-like reptiles. This transition, which was originally thought to have been a rare, unrepeatable, contingent event, seems to have happened twice in the early mammalian lineage based upon the evidence in the fossil record.[30]

Other traits are even more convergent. Ovoviviparity, in which the egg is retained within the female reproductive track prior to a live birth, has evolved over one hundred times in lineages as diverse as amphibians, reptiles, and fish. While ovoviviparity is distinct from mammalian development in utero, many of these non-mammals even have placental-like structures by which they can nurture their young. Another key mammalian trait, endothermy (warm-bloodedness), has evolved multiple times in mammals, birds, fish, and insects. The point of this is that the mammalian traits are not highly contingent, one-time occurrences, evolutionarily speaking.[31] Rerun the process again and you are likely to get creatures with these traits. While the pattern and timing of the emergence of these traits would be

30. T.H. Rich et al., "Independent Origins of Middle Ear Bones in Monotremes and Therians," *Science* 207 (2005): 910–914.

31. Conway Morris, *Life's Solution*, 218–228.

different each time due to the contingent aspects of evolution, these solutions represent such good ways of "making a living" on our planet that, given enough time, the likelihood that they would emerge again is quite high.

Given the massive amount of convergence surrounding the evolution of mammalian traits, the interpretation that the K-T event was absolutely necessary for the mammals to arrive as the dominant predators on the planet seems suspect. In addition, some larger mammals that likely preyed upon dinosaurs already existed prior to the K-T event.[32] While they were not as diverse as the Paleogene mammals, the Cretaceous mammals had already diversified into a vast array of niches before the K-T event. The fact that the mammals were able to regulate their body temperature and safely maintain their young internally before birth would have allowed them to exploit a diverse range of niches other groups could not (particularly in terms of surviving colder temperatures). This ability would have provided the group with an advantage in the long run, particularly on a planet that periodically undergoes massive environmental perturbations.

The point is that, in a constantly changing environment, it appears that the mammalian form is more resilient/advantageous than that of the dinosaurs. If the K-T event had not brought the dinosaurs down, is it not likely that another of the major environmental upheavals that regularly occur on the planet would have? Conway Morris posits that "as the Tertiary planet cooled [the Tertiary is the geological period from sixty-five million years ago to the present], culminating of course in the ice ages, the mammals . . . would have been at an advantage anyway, well adapted to temperate and even polar environments. This suggests that the rise of active, agile, and arboreal ape-like mammals, would have been postponed [without the K-T event], not cancelled."[33]

32. G. Wilson et al., "A Large Carnivorous Mammal from the Late Cretaceous and the North American Origin of Marsupials," *Nature Communications* 7, no. 13734 (2016), https://doi.org/10.1038/ncomms13734.

33. Conway Morris, *Life's Solution*, 222.

His position is that evolution, though messy in the day-to-day details, ultimately trends toward certain predictable solutions. Particular life forms or traits are hardwired into nature, woven into the very chemical and physical fabric of our universe. They are such good solutions that the evolutionary process cannot help but eventually find them. It is attracted toward them like a ball careening down a rocky hillside is attracted toward the bottom of the hill, its path unpredictable but its destination certain.

CHANCE, PURPOSE, AND GOD'S PROVIDENTIAL PLAN

It is hard to dismiss the striking level of convergence seen in evolution. It is also hard to dismiss the fact that chance events do influence evolutionary trajectories such that their paths are not readily predictable. This is the reality of the evolutionary process: it is a balance of chance and order, of creativity and constraint. It is an inherently chancy process, but one that is being played out upon a highly structured bed of order.

How much does this underlying order restrict or limit what these chance processes can uncover? To what extent are evolutionary trajectories dictated by chance events? Is our universe ordered to the extent that the emergence of humankind is inevitable once the evolutionary process is set in motion? Simon Conway Morris argues strongly for this position, but it may turn out that, despite the high degree of evolutionary convergence, we owe our existence as a species to certain chance events in the evolutionary past, events without which a human-like species would not have emerged. Yet, even if this turns out to be the case, it does not mean that the origin of mankind falls outside of God's providential care. Chance events really do influence our lives and really do influence the evolutionary process. But a God who sustains creation at every moment, one who allows created things to act as causes in their own right, also sustains all the chance encounters that occur among created causes during the evolutionary process.

Those chance events that we actually observe in evolution *are* his plan, although from *God's* atemporal perspective, it's hard to call such events "chance" at all. The philosopher Ernan McMullin drives home this point well:

> It makes no difference . . . whether the appearance of *Homo sapiens* is the inevitable result of a steady process of complexification stretching out over billions of years, or whether on the contrary it comes through a series of coincidences [chance events] that would have made it entirely unpredictable from the (causal) human standpoint. Either way, the outcome is God's making, and from the biblical standpoint could be properly said to be God's plan. Terms like 'plan' obviously shift meaning when the element of time is absent. For God to plan is for the outcome to occur. There is no interval between the decision and the completion.[34]

As important as it is to be able to reconcile chance and providence, it is equally important to not let the reality of chance events obscure the more important fact that our universe is highly ordered, particularly at its most foundational level. Science has made incredible progress toward describing this order and understanding how order at one level, at the level of small molecules for instance, influences the emergence of order at higher levels, such as in protein structure. Despite this success, there is a foundational level of order that scientists will be able to describe that *cannot* be explained by appealing to some lower or more fundamental level. This foundational order just *is*. While science can measure it, describe it, and model it, it cannot explain *why* we live in a universe with this particular foundational order. What then is the source of this order? From a Catholic perspective, this

34. Ernan McMullin, "Cosmic Purpose and the Contingency of Human Evolution," 1996, Pari Center website, https://paricenter.com/library/culture-philosophy-and-science/cosmic-purpose-and-the-contingency-of-human-evolution/.

foundational order that science can describe is best explained as originating from the Author of nature, the first cause of all that is.

One must be clear that such a conclusion is philosophical and not scientific in nature. However, it is one that is entirely consistent with the scientific data regarding evolution. In fact, far from drawing one away from faith, an understanding of the evolutionary process, given the order that sustains and influences it, represents one more point of encounter with the Creator, the ultimate source of the order that runs through every level of our universe.

7

Evolution and Aquinas

Emergence, Change, and the Potentiality of Matter

All these [living] things around us have been seminally and primordially created in the very fabric, as it were, or texture of the elements; but they require the right occasion actually to emerge into being.[1]

—St. Augustine

Thomas Aquinas, the Angelic Doctor, quite rightly stands out as one of the most influential philosophers within the Catholic tradition. His overall body of work, in particular his *Summa theologiae* and *Summa contra Gentiles*, has shaped the development of Catholic teaching and doctrine for the past eight hundred years. Given his tremendous influence on Church teaching, many modern Catholic Aquinas scholars have attempted to examine the implications of evolutionary theory through the lens of Aquinas' metaphysics.[2] In large part, these writers have sought to reconcile modern evolutionary thought with Aquinas' views on creation, nature, and change. For example, Jacques Maritain, one of the leading Thomists of the twentieth century, wrote an extensive essay entitled "Towards a Thomist View of Evolution" that discussed, among other topics, how the evolutionary origin of the first humans might fit within a Thomistic framework. Using distinctive Thomistic language, Maritain posited that an advanced hominin[3] might have been able to produce an offspring

1. Augustine, *The Trinity* 3.16, trans. Edmund Hill, OP (Brooklyn: New City, 1991), 137.

2. This list includes a variety of Thomists, such as Jacques Maritain, William Carroll, Mariusz Tabaczek, Ernan McMullin, and Antonio Moreno.

3. Hominins are fossil species that are more similar to modern humans than modern chimps.

in which the fusion of a modified sperm and/or egg could result in an entity that had the ultimate disposition to be informed by a rational human soul (the human form):

> This ultimate disposition takes place at an *instant of time*, at a single and identical chronological instant when the form that has been informing up till that moment—the sensitive or animal soul of the hominian fetus in the present case (and we should recall that this happens likewise in the case of human embryonic development)—returns into the potency of matter. At this same instant the newly informing form is either educed from the potency of matter, or else—in the present case [the origin of man]—not educed from the potency of matter but rather created by God and infused into the fetus.[4]

As he indicates in the above passage, Maritain's view that advanced hominins could produce the type of offspring capable of being informed by a human soul finds its roots in Thomas' view of normal human development. Thomas argued that rationally ensouled humans do not emerge until forty days post-conception for boys and eighty days post-conception for girls. While Thomas' view is no longer accepted, his rationale for this position is founded upon a type of "evolutionary" transition occurring *in utero*. In his view, the early developing fetus was not the appropriate type of entity (it lacked a developed brain, for example) in which the substantial from of a rational human could be actualized. Rather, at this early stage the human form was present only virtually and "emerged" later on in development. In the *Summa contra Gentiles*, Aquinas discusses how this might operate:

> Thus, prime matter is in potency, first of all, to the form of

4. Jacques Maritain, *Untrammeled Approaches*, in *Opera Omnia*, trans. B.E. Doering (Notre Dame, IN: University of Notre Dame Press, 1997), 10: 118–129, https://inters.org/maritain-dynamism-nature-evolution.

> an element. When it is existing under the form of an element it is in potency to the form of a mixed body, that is why the elements are matter for the mixed body. Considered under the form of a mixed body, it is in potency to a vegetative soul, for this sort of soul is the act of a body. In turn, the vegetative soul is in potency to a sensitive soul, and a sensitive one to an intellectual one. This the process of generation shows: at the start of generation there is the embryo living with plant life, later with animal life, and finally with human life.[5]

This discussion of human development in which the fetus has a vegetative, then sensitive, and finally rational soul describes how one form, as it develops, has the potency to be informed by another higher form. For Aquinas, one soul did not evolve into another type of soul during this process. Rather, the vegetative soul "returns into the potency of matter" such that the matter can then be informed by the sensitive soul and have distinct properties and abilities.

While Aquinas does not extend this in evolutionary time to envision one organism giving rise to another type of organism, he does indicate in this passage the tendency of one form to give way to a higher form. If, according to Aquinas, such an upward movement of forms occurs in human development, would there be anything within his metaphysics to preclude there being an inherent upward tendency toward more complex and higher forms appearing over time in nature? If the movement from sensitive (animal) soul to intellectual soul can take place in the matter of days *in utero*, what is there to prohibit the possibility of much smaller transitions, one type of animal giving rise to matter—a materially novel fertilized egg—that can be informed by another type of animal soul, within Thomas' metaphysics?

While Maritain and others believe that such transitions are

5. Thomas Aquinas, *Summa contra Gentiles* 3.22.7, trans. Vernon J. Bourke (Notre Dame, IN: University of Notre Dame Press, 2001), 86.

in principle consistent with Thomas' metaphysics, there are some Catholics who believe that Thomistic philosophy is incompatible with this idea. Furthermore, some have even argued that Aquinas' work demonstrates that macroevolutionary transitions are impossible, and that to hold such evolutionary views contradicts not only Aquinas' position but fundamental Church teachings as well.[6]

To make their argument, these critics regularly point to the many passages in Aquinas' writings in which he argues that biological species were created *directly* by God. For example, in his *Commentary on the Sentences*, book 2, he states, "[Some things come into being neither through motion nor through generation] because of the necessity that generation always generates what is similar in species. For this reason, the first members of the species were immediately created by God, such as the first man, the first lion, and so forth."[7] The fact that Aquinas believed that higher animal species such as lions and men were created immediately by God—and that he stated this multiple times in his works—is not surprising given that the direct creation of such higher species was the dominant view of the time. In fact, no medieval thinker, given the predominant scientific views of the period, had any reason to develop an evolutionary account of the origin of such species.

Yet Aquinas' view on the origin of species was much more complex. In fact, Aquinas was open to the possibility that new species could indeed have come into existence via natural processes after the six days of creation. He was open to this position based upon his actual experience of the natural world, as he articulates in the *Summa*:

6. Both *Aquinas and Evolution* (Leicester: Chartwell, 2017) and *Catholicism and Evolution: A History from Darwin to Pope Francis* (Kettering, OH: Angelico, 2015) by Fr. Michael Chaberek as well as *The Metaphysics of Evolution: Evolutionary Theory in Light of First Principles* (Sensus Traditionis, 2012) by Fr. Chad Ripperger aptly illustrate this type of argumentation.

7. Thomas Aquinas, *In II Sent.* 1.1.4, in *Aquinas on Creation: Writings on the "Sentences" of Peter Lombard 2.1.1*, trans. Steven Baldner and William Carroll (Toronto: Pontifical Institute of Mediaeval Studies, 1997), 85.

> Species, also, that are new, if any such appear, existed beforehand in various active powers: so that animals, and perhaps even new species of animals, are produced by putrefaction by the power which the stars and elements received at the beginning. Again, animals of new kinds arise occasionally from the connection of individuals belonging to different species, as the mule is the offspring of an ass and a mare: but even these existed previously in their causes, in the works of the six days.[8]

Here Aquinas invokes the possibility that new species could emerge over time, but their potentiality must have been there from the beginning. Just as he argued that the matter of an element lies in potency to a compound, the matter of some species (the ass and the mare) exists in potency to other species (the mule).[9] In fact, this passage seems to mirror in some respects Augustine's concept of the "seeds" or "order" of new species being planted in creation at the beginning. These seeds then allow the emergence of species at their proper time in accord with their proper causes.

Aquinas even discusses how plants would have been produced in this fashion, emerging from the interaction of the earth and the heavens: "Hence the plants were not actually produced on the third day but only in their causes: and after the six days they were brought into actual existence in their respective species and natures by the work of government."[10] As such, Aquinas' writing does seem to leave open the possibility of evolutionary transitions given that he is open to the emergence of new species instrumentally from material causes (the earth and the heavens). Yet at the same time, he clearly argues that some new species are created

8. Thomas Aquinas, *Summa theologiae* 1.73.1 ad. 3.

9. The potency in matter for certain forms to evolve is something that is discussed in chapter 5 and later in this chapter regarding evolutionary convergence. For example, the protein folds that have evolved are dictated largely by the underlying chemistry, which favors certain secondary structures and excludes other possibilities. In a very real sense, the living forms that exist today existed in the potency of matter at the dawn of the universe.

10. Aquinas, *De Potentia Dei* 4.2 ad. 28, vol. 2, trans. the English Dominican Fathers (Westminster, MD: The Newman Press, 1952), 63.

immediately by God. To begin to make sense of this apparent dichotomy, it is useful to start with a brief overview of Aquinas' metaphysics.

AQUINAS AND CREATION

One oft-repeated Thomistic critique of evolution is the fact that Aquinas reserves the act of creation to God alone. In the *Summa*, Aquinas states, "Now to produce being absolutely, not as this or that being, belongs to creation. Hence it is manifest that creation is the proper act of God alone. . . . Therefore, it is impossible for any creature to create, either by its own power or instrumentally."[11] Based upon this principle, Thomistic philosophy would seem incompatible with the notion that God used the evolutionary process instrumentally to create new species. Rather, as Thomas argued, creation is the proper direct act of God alone.

This apparent contradiction, though, hinges on a distinction between what Aquinas means by the term creation and the common meaning of that term. In popular parlance, the term creation often is used to refer to the origin of something, the time when it happened to be fashioned. For example, we say an artist creates a picture using paint and brushes, or a baker creates a cake using eggs, flour, and sugar. While this usage of the term creation is quite common, Aquinas has a much more specific usage of the word. For Aquinas, the examples of the artist and the baker "creating" a picture or a cake would not, strictly speaking, be seen as acts of creation. Rather, he would classify them as examples of *change*. For example, the baker takes certain proportions of ingredients and heats them such that they *change* into a cake.

Creation, for Aquinas, is a metaphysical principle that means something very specific: the radical dependence everything that exists has on God for its very existence. (This notion was discussed in detail in chapter 3.) In addition to creating the

11. *ST* 1.45.5 co.

universe out of nothing at the very beginning, God sustains all of creation at *each and every moment*. All created entities (rocks, frogs, hydrogen atoms) are contingent entities in that they do not exist out of necessity. Furthermore, they cannot bring themselves into existence, as none of these contingent entities contains the power of existence within itself. Rather, it is God who brings and sustains them in existence at each and every moment they exist. As Aquinas argues in the *Summa*, "The being of every creature [the *esse*] depends on God, so that not for a moment could it subsist, but would fall into nothingness were it not kept in being by the operation of the Divine power."[12]

The mere fact that something exists, or as Aquinas says, is "kept in being," is entirely dependent upon God, the ultimate source of existence. However, the manner by which something is *brought into existence* can for Aquinas involve secondary causes such as a baker or an artist. In the case of a cake, the baker and the oven are secondary causes that operate to bring the cake into existence. These causes, through properly mixing the ingredients and providing the proper temperature, allow the ingredients to take on the properties that we recognize as a cake. The cake does not spring into being *ex nihilo* but is the result of a material change. Likewise, physical evolutionary transitions that bring about new species are the result of material changes—not creation in the sense used by Aquinas. The Dominican philosopher and theologian Fr. Mariusz Tabaczek explains this distinction as follows: "Even though Aquinas clearly states that *esse* [existence] has its ultimate source and can only be 'produced' by God, he admits that creatures can be causes of coming into existence of other created entities. As such, they may be called causes but not of existence (*esse*) as such (i.e., *causa essendi*), but of coming into existence (i.e., *causa fiendi*). In other words, they

12. *ST* 1.104.1 co.

may be called secondary causes of coming into existence (acting under the primary causation of God)."[13]

This principle in Aquinas, the ability of creatures to be causes of the coming into existence of other created entities, leaves open the possibility for the evolutionary origins of species. However, the fact that anything exists at all is dependent upon God alone as Creator. No other cause can bring things into existence out of nothing or sustain things in existence. However, secondary causes can *change* pre-existing entities into new entities. In fact, this occurs without excluding the unique creative role of God. While the baker and the oven may be the efficient causes of the cake changing into its currently recognizable structure of a cake, this does not exclude the causative powers of God who sustains both their *esse* and the *esse* of the cake. Within Thomistic metaphysics, to give existence to something is the primary act of creation, an act that is reserved for God alone, for he alone is pure *esse*.

Another important distinction that Aquinas makes in his metaphysics is between form and matter. For Aquinas, all material substances that exist in the universe are composed of a unity of matter, substantial form, and accidental forms. Prime matter is the metaphysical principle of pure potentiality. The substantial forms actualize this prime matter to give a substance its individuality as a particular type of thing. For example, a squirrel has the substantial form of a squirrel that actualizes the prime matter into this particular form and makes it recognizable as a squirrel. Each squirrel, though, is also informed by accidental forms, which simply modify some existing substance, in this case the substance of a squirrel. These accidental characteristics, such as size, hair color, and tail length, can change without disrupting the underlying substantial form of squirrel.

For living substances, Aquinas refers to the substantial form

13. Mariusz Tabaczek, "What Do God and Creatures Really Do in Evolutionary Change? Divine Concurrence and Transformism from the Thomistic Perspective," *American Catholic Philosophical Quarterly* 93, no. 3 (2019): 458–459, https://doi.org/10.5840/acpq2019514179.

as the soul, the animating principle of the living organism. He classifies three types of souls and therefore three types of living organisms. Plants have a vegetative soul, which endows them with the powers of nutrition, growth, and reproduction. Animals have sentient souls, which endow them with the additional properties of sensation and movement. And humans have rational souls, which endow them with all these properties plus the ability for rational thought.

Aquinas argued that all of these properties, with the exception of rational thought, were material in nature. Given that the powers of animal and vegetative souls are completely reliant on the matter of the organism, he considered them corruptible and unable to survive the death of the organism. The human rational soul, on the other hand, endows humans with the immaterial power of rational thought. Because of this property, which for Aquinas is not grounded in the matter of the organism, the human soul is considered both immaterial and immortal (incorruptible). He further argued that the human soul must therefore be created by God immediately (in accordance with normal biological reproduction) and is not educed in some form from the arrangement of the matter that makes up the organism.[14]

The question then regarding evolution is how and whether one organism with a specific substantial form (soul) can give rise to another type of organism with a distinct substantial form. What might be able to cause this type of substantial change within Aquinas' metaphysics? Except for the case of the origin of humans, in which the creation of an immaterial rational soul is necessary, such evolutionary changes are not properly viewed as creation events in the sense that Aquinas uses the term. Rather, evolutionary *change* consists of one substantial form actualizing prime matter in such a way that it is disposed to being informed by a different type of substantial form.

14. *ST* 1.75.6.

HOW DO NEW SPECIES COME INTO EXISTENCE WITHIN A THOMISTIC FRAMEWORK?

In the *Summa*, Thomas indicates in certain passages that the first instance of a species must be created directly by God. For example, he states that "the first production of corporeal creatures is by creation, by which matter itself is produced . . . and it is impossible that anything should be created, save by God alone."[15] The reason for this is that Aquinas is assuming the production of the first instance of a species occurs *ex nihilo*, as he associates the "first production" with matter being produced. Later in the same passage, he further explains his rationale, stating that "creation is the production of a thing in its entire substance, nothing being presupposed either uncreated or created. Hence it remains that nothing can create except God alone."[16] However, in an evolutionary transition, we do presuppose things, particularly the existence of other substances (organisms) that contribute matter that can be informed by a different substantial form.

The question then is not whether an evolutionary process can create something out of nothing (it clearly cannot, and Aquinas is correct on this point) but whether it can bring about a *change* such that a new form is educed or brought out of the potency of prime matter. How might this be explained? Using the same Thomistic framework discussed above, biological organisms, along with a plethora of environmental factors, could be seen as interacting within a complex network of secondary and instrumental causes in order to bring into existence, out of the potential inherent in prime matter, new forms. These new forms, because they share in God's existence, are also immediately created by God in the sense that God immediately creates and sustains everything that exists at each and every moment. Using distinctly Thomistic language, Tabaczek summarizes this position as follows: "Paraphrasing

15. *ST* 1.65.3 co.

16. *ST* 1.65.3 co.

Aquinas's assertion from *ST* I, q. 65, a. 4, co. we might say that the corporeal form (as such) that the first exemplar of the species has when first produced comes immediately from God, while its eduction from the potentiality of primary matter effects from the secondary and instrumental causality of other creatures [as well as the environmental factors present]."[17]

He goes on to state that "evolutionary change is neither simply generation [change], nor a direct creation *ex nihilo* (which in Aquinas's interpretation of the work of six days seems to refer to the most basic elements)."[18] Rather, from a Thomistic perspective, one must integrate these two principles, principles that undergird all change in the created world. Things that come into being are generated via change through the operation of secondary and instrumental causes, but they are also created directly by God because God's creative power as the ultimate source of existence "keeps in being" all contingent entities at all times.

To fixate on passages from Thomas in which he attributes the origin of certain new living substances to direct creation misses this key point. In addition, it does not address the fact that Aquinas argues in other passages that the first instances of some species such as plants or lower animals do not need to be created directly by God. In fact, in certain passages Aquinas agrees with Augustine that the earth was given the ability or power to produce plants, and in others he agrees with Aristotle that certain animals can be produced via putrefaction from nonliving matter.

Regardless of this issue, Aquinas does make a clear distinction between the powers of creatures to act as causes of species and the ultimate causation of God. He points out that creaturely causation, which produces entities through change (motion) and reproduction (generation), is subordinate to God's ultimate causation. While other created beings can have a role in the production of lions and horses, Aquinas states that "God is also the cause

17. Tabaczek, "What Do God and Creatures Really Do in Evolutionary Change?"
18. Tabaczek.

of these things, operating more intimately in them than do the other causes that involve motion, because He Himself gives being to things. The other causes, in contrast, are the causes that, as it were, specify that being."[19] If one applies this reasoning to the origin of new species, one could argue within a Thomistic framework that the variable matter contributed by the parents in the sperm and oocyte (along with the environmental conditions and influences that are present upon that zygote) can specify or be determinate of a new type/species of being coming into existence. However, one must bear in mind that the fact that this new being exists at all is dependent solely upon God.

While Aquinas seemed confident that species did not change or evolve and that the first lion, for example, could not have come from any other created species, his position here does not seem to be driven primarily by his philosophical commitments. Rather, it seems to flow out of the scientific understanding of his time. Had the scientific understanding of the time been different, it is not out of the realm of possibility that his careful distinction between the ability of creaturely causes to effect change and God's ultimate role as the primary cause of all that exists would have allowed him to integrate the evolutionary origins of species into his philosophical system.

MORE PERFECT FROM THE LESS PERFECT

Aquinas' distinctions between primary and secondary causality, between substance and form, and between creation and change appear to leave room for integrating evolutionary transitions into his philosophical system. However, there are other aspects of Aquinas' philosophy that raise concerns regarding the compatibility of his metaphysics with evolution. One such issue involves

19. Thomas Aquinas, *In II Sent.* 1.1.4, in *Aquinas on Creation: Writings on the "Sentences" of Peter Lombard 2.1.1*, trans. Steven Baldner and William Carroll (Toronto: Pontifical Institute of Mediaeval Studies, 1997), 85.

a variation of the Scholastic principle of sufficient reason. While this principle has been articulated in a variety of manners, for our purposes it can be stated as the following: 1) whatever comes to be must have a cause of its coming to be, and 2) that cause must be adequate to produce it. In the context of evolution, the key issue is that the cause must be sufficient to produce the effect. If a simpler organism is thought to eventually give rise to a more complex organism during an evolutionary transition, this seems to violate that principle. It would seem that the evolutionary "lower" or less perfect organism would lack the ability within itself to be cause of the "higher" or more perfect organism. For example, a squirrel cannot give rise to a chimpanzee, nor can a chimpanzee give rise to a human. Rather, squirrels only have the power to give rise to other squirrels via reproduction. Because they possess only a squirrel substantial form, squirrels lack the ability to directly give rise to any "higher" or more "perfect" organismal forms. This principle is often succinctly summarized by the phrase "one cannot give what one does not have."

However, according to the Thomistic philosopher Brian Carl, this argument is based upon an oversimplification of Aquinas' view on reproduction. For Aquinas, an individual squirrel does not, on its own, have the power to produce new squirrels. As Carl explains,

> It is not in fact Thomas's view that the individual generating animal is on its own a sufficient agent cause of the generation of a new animal within its own species, since the generating animal depends upon its being moved by a higher agency in order to cause a thing's nature. The individual animal does not have its nature, on its own, in such a manner that it is able to give that nature to something else; it has that nature in such a manner that it can be used instrumentally by a superior cause.[20]

20. Brian Carl, "Thomas Aquinas on the Proportionate Causes of Living Species," *Scientia et Fides* 8, no. 2 (2020): 235.

In Aquinas' philosophy, this superior cause in the generation of animals is to be found in the heavens. It is not the squirrel that educes the substantial form, but rather the power of the heavens, working through the squirrel, that has this ability. For Aquinas, the sun is the superior cause, one that uses instrumental causes such as the male squirrel's semen to educe the form of a new squirrel from the potency of the matter that is provided during the process of squirrel reproduction. Squirrels do not have the power to produce squirrels on their own, but rather the sun's energy is the "superior cause" that is able to bring about the new squirrel, and it does this through the instrumental use of the parental squirrels (particularly, for Aquinas, the father's semen).

In fact, the heat of the sun is so powerful that it can educe the form of some animals without the use of an instrumental cause. In his discussion of spontaneous generation,[21] Thomas argued that the sun could produce simple or lower animals without the focusing work of the semen. These organisms could be produced spontaneously and arise via chance interactions based upon how the matter in the slime just so happened to be arranged at the time it was heated by the sun. In a sense, the semen focused the power of the sun to produce a specific form, while in the case of spontaneous generation, the power of the sun was left to produce whatever form could be educed from the arrangement of matter that just so happened to be there in the slime. While Aquinas clearly had his biology wrong, the critical point here is Aquinas' view on reproduction—namely, that the organism does not have the power to educe the form of its offspring on its own; rather, the form is educed or drawn forth by a higher cause (the sun).

One of the reasons that Aquinas, like Aristotle before him, believed that the sun had a role in the generation of new organisms is that he believed that whatever the cause of the essence of a specific animal, it had to lie beyond any individual member

21. Thomas, like Aristotle before him, believed that organisms could emerge as the sun putrefied flesh or slimy mud.

of that species. For example, a horse, because it participates in the substantial form of "horseness," could not, on its own, be the cause of "horseness." According to Aquinas, "Since then, the generating horse has the same [specific] nature, it would have to be its own cause, which is impossible. It remains, therefore, that above all those participating in equinity [horseness], there must be some universal cause of the whole species."[22]

In search of a universal cause of form, Aquinas sought to locate it somewhere in the material world. Unlike Plato, who located it in the transcendent world of universal forms, Aquinas located the universal cause within the natural world, in the perfection of the heavenly spheres. Thus, the form came from something beyond the organism yet something that was firmly entrenched within the material world.[23]

Of course, this is not how reproduction works, but the point in dispute here is not the accuracy of his biology but the amenability of his metaphysics to the evolutionary origins of species. As Steven Baldner and William Carroll have pointed out, "Aquinas is saying that animal and plant generation need not, in principle, always take place from parent members of the species."[24] (This is particularly evident in Aquinas' views on putrefaction.) Because the causal powers of animal and plant generation are not solely located in the parents within Aquinas' system, it therefore does not preclude the possibility of the sun's power using an organism as an instrumental cause to produce a novel form. This novel form would be the result of the matter being disposed differently—for example, the sperm being altered—such that the higher power of the sun educes a "higher" form that is novel.

22. Thomas Aquinas, *De substantiis separatis* 10.58, trans. Francis Lescoe (Carthagen, OH: The Messenger, 1963), 96.

23. Aquinas also rooted the form in the mind of God. For example, he speaks of divine ideas as exemplar causes given that the form of horse, and every other form for that matter, originated in the mind of God. He compares this to how the intellect of the artist contains and is responsible for all the works of the artist. However, the bringing about of a particular substantial form in the world relied on the perfection of the heavenly spheres.

24. Thomas Aquinas, *Aquinas on Creation,* 85n51.

The secondary cause of this novel substantial form would not be merely the "simpler" organism, but rather the superior cause of the sun.

It is interesting to note that Aquinas' position—that the form is educed from causes that extend beyond the parents—finds a parallel in modern evolutionary biology. New research over the past few decades has demonstrated that the emergence of a new species can involve multiple factors that go beyond the genetic information of the parental species. This new understanding, which is referred to as the extended evolutionary synthesis (EES) and is discussed in more detail in chapter 4, does not negate the role of natural selection, but rather expands our understanding of evolutionary processes. For example, the physical and environmental constraints discussed in chapter 5—along with a complex nexus of feedback loops involving environmental factors, ecosystems, developmental programs, organismal behavior, and species interactions—all play critical roles in shaping the emergence of novel biological species. This multitiered approach is described in a recent article penned by a number of biologists at the center of the EES movement: "In developmental processes that generate biological form, for instance, cellular architecture, tissue activity, physiological regulation, and gene activation play together in intricate functional networks, without any privileged level of control. Evolutionary modification of such multilevel dynamics, be it through mutation, natural selection, or environmental induction, will always affect the entire system. By necessity, such multilevel systems exhibit emergent properties."[25]

Our current understanding of the causal factors involved in the origin of new species is much richer than the outdated gene-centric approach in which evolution could simply be reduced to genetic changes in the gametes. This new understanding involves a dynamic interaction between complex factors at the

25. Denis Noble et al., "Evolution Evolves: Physiology Returns to Centre Stage," *Journal of Physiology* 592, no. 11 (2014): 2237–2244.

level of the gene, the organism, and the ecosystem, all of which are influenced by physical and chemical laws and constraints. Therefore, arguing that species *x* cannot produce species *y* merely because species *y* is more complex fails to take into account both Aquinas' actual views on reproduction and recent developments within the science of evolution.

FORMS IN BIOLOGY AND THOMISTIC PHILOSOPHY

One possible confusion that emerges when trying to integrate evolutionary biology with Thomistic metaphysics is the use of the term *form*. As described previously, within Thomistic philosophy the form of a substance is the principle that makes a thing what it is. It is the principle that informs prime matter, that indefinable potential of being, such that it becomes something we can recognize as an organized whole. However, when biologists use the term *form*, they use it with quite a different meaning. Biological form refers to the physical structure and organization of an organism that can be measured and quantified. For the philosopher, each dog possesses a substantial form of "dogness" such that we can recognize it as a dog. For the biologist, each dog has a particular biological form that can be measured and quantified and investigated scientifically.

While biologists and philosophers use the term differently, there are interesting parallels between the way Thomistic philosophers and modern biologists use the term. For instance, Thomistic philosophers often refer to the form as being educed from the inherent potency in prime matter. Aquinas himself repeatedly discusses how one substance "is in potency" to becoming something else, prime matter informed by a different substantial form. Interestingly, a similar concept seems to be at work when one examines the emergence of biological forms. In a very real sense, the biological forms we see around us have emerged from the potency of the properties inherent in physical matter. (This is of course different from the philosophical concept of prime matter.) In other words, because of the way the universe is physically structured, certain chemical and biological forms will emerge over evolutionary time, while others simply have no possibility of coming into existence. Furthermore, the physical forms that do come to exist emerge not merely from some cosmic accident; rather, they tend to emerge because they represent robust and stable solutions that flow out of the specific physical order that undergirds our universe. For example, camera-like eyes have emerged multiple times during evolution on our planet, presumably because such a sense organ is a good solution given 1) the properties of light in our universe and 2) the survival advantage of sensing and focusing this light. On the flip side, elephant-sized insects have never emerged because this is not a viable biological form given the physical constraints of having an exoskeleton.[26]

Another rather illustrative example of this concept is seen with the physical form of biological proteins. As discussed in chapter 6, proteins generally fold into only two stable secondary structures: alpha-helices and beta-sheets. The reason that these structures tend to be stable can be traced back to the underlying chemistry and physics. In proteins, the bond angles between the connected atoms—the charges associated with the atoms—and the hydrophobic/hydrophilic interactions between the bonds and atoms dictate that alpha-helices and beta-sheets are the stable forms that emerge when amino acids are strung together in proteins.

Furthermore, these stable secondary structures tend to be packed together based on specified rules that limit the arrangement of their orientations to a relatively small number of protein forms or folds.[27] As such, there seems to be only a limited number

26. This topic is discussed in more detail in chapter 6.

27. The manner in which they interact to form three-dimensional proteins is dictated and limited by chirality preferences, symmetry considerations, steric constraints, and the interactions between hydrophobic and hydrophilic residues within the protein. See C. Chothia, "Principles that Determine the Structure of Proteins," *Annual Review of Biochemistry* 53 (1984):

of protein forms that are available to nature, a situation that is dictated not by evolution but by the underlying properties of matter and the underlying fundamental forces.

In a very real sense, the protein forms that exist in nature emerge from the potentiality inherent in the underlying structure of the universe. If the universe had been constructed using a *different* set of physical laws, a *different* set of potentialities would likely exist and a different set of protein forms would be made manifest by evolution. The biochemist Michael Denton likens the forms of proteins to attractors, functional possibilities toward which amino acid chains are drawn by physical and chemical rules. In fact, he goes so far as to liken the limited number of biological folds found in proteins to Platonic forms, somewhat blurring the distinction between biological and philosophical explanations.[28]

While Denton takes it further than most biologists, there is now a widespread appreciation among biologists for the role that physical laws/constraints have had on the evolution of biological form. This applies not only at the level of protein folds but also to higher-level structures as well. For example, the fractal branching pattern of blood vessels is largely the result of physical constraints. Based on considerations of fluid dynamics and diffusion (generic laws that are shared with nonliving entities), a fractal branching pattern is the only efficient way to distribute or collect fluids from a central source (for example, the heart) and circulate to all the cells throughout the body. In fact, this pattern is seen repeatedly throughout nature, from the root system of trees to the bronchiole system in mammalian lungs to the tracheal system in insects.

Likewise, the streamlined spindle-shape of aquatic organisms is found in diverse branches of life, such as bony fish,

537–572; M.J. Denton et al., "Physical Law Not Natural Selection as the Major Determinant of Biological Complexity in the Subcellular Realm: New Support for the Pre-Darwinian Conception of Evolution by Natural Law," *Biosystems* 71 (2003): 297–303.

28. Denton et al., "Physical Law Not Natural Selection."

cartilaginous fish, and mammals. Given the physics of moving through a viscous watery fluid, a feature that is set by the physical structure of the cosmos, the streamlined spindle-shape form is one that the evolutionary process will repeatedly discover. Other examples of the physics dictating form can be seen with eye development, the origin of powered flight, and body size to metabolism ratios in mammals. Furthermore, the physical dynamics of nonlinear systems influence large aspects of biological development.[29] In such systems, small perturbations can have large effects, as seen in the case of limb development. In this case, one small genetic change can produce a limb with an additional bone, but one that integrates properly within the physiology of the limb, complete with the necessary muscular connections and vasculature.[30]

According to the biologist Stuart Newman, this connection changes the manner by which one views the evolutionary process: "If physics plays a causal role, along with genes, in the development of organismal form, there must be aspects of development and its outcomes . . . that are 'generic' and are explicable in terms of processes that living systems have in common with non-living ones."[31] In fact, the whole principle of convergence that was discussed in chapter 6 is based upon this realization—namely, that certain biological forms emerge, often repeatedly, not merely due to chance but also due to these biological forms being the robustly favored possibilities (potentialities) given the physical fabric of the cosmos.

Clearly there is a deep and necessary connection between physics, chemistry, and living organisms. While this observation may seem trivial (the chemistry and physics must be amenable to life since life exists), from it flows a profound realization. The

29. Stuart A. Newman, "Form and Function Remixed: Developmental Physiology in the Evolution of Vertebrate Body Plans," *Journal of Physiology* 592, no. 11 (2014): 2403–2412.

30. M.B. Hawkins et al., "Latent Developmental Potential to Form Limb-like Skeletal Structures in Zebrafish," *Cell* 184, no. 4 (2021): 899–911.

31. Newman, "Form and Function Remixed."

cosmos is structured in such a manner that it had the potential for producing all the living forms we see around us from the very beginning. The evolutionary process, both cosmically and biologically, gradually prepared matter to reveal this latent potentiality, a potentiality shaped by the structure of the cosmos. While this is a purely scientific claim, it is strikingly analogous (but not identical) to the Thomistic philosophical claim that new substantial forms can be educed from the potentiality of matter.

Aquinas discusses how simpler substances lie "in potency" and set the stage for more complex substances. He states that elements have the potency to form mixed bodies, for example. Again, there is an analogy here to evolution, as each stage of evolutionary development sets the foundation for the next. The emergence of the simpler prokaryotic cells sets the stage for the emergence of more complex eukaryotic cells, which sets the stage for the emergence of multicellular organisms, and so on.

Aquinas talks about how the substantial forms that exist around us flow from God as the "first exemplar cause of all things."[32] For Aquinas, these forms flow from the mind of God in the manner in which the form of the statute of David flowed from the mind of Michelangelo: "For an artificer produces a determinate form in matter by reason of the exemplar before him, whether it is the exemplar beheld externally, or the exemplar interiorly conceived in the mind. Now it is manifest that things made by nature receive determinate forms. This determination of forms must be reduced to the divine wisdom as its first principle, for divine wisdom devised the order of the universe."[33]

Aquinas' notion of God as the exemplar first cause resonates with the notion that the order of the universe, devised by divine wisdom, directs the evolutionary process toward specific forms. God could have created any universe he wanted, but he chose to bring forth this specific universe with this specific underlying

32. *ST* 1.44.3 co.

33. *ST* 1.44.3 co.

order. All of the magnificent forms that we see around us have emanated from the mind of God over cosmic time via a spectacularly rich interplay between order and chance, *in just the manner he conceived.* And it happened not in some deterministic fashion (although the underlying order has certainly driven evolution toward certain solutions), but rather in a creative fashion in which the potentiality of the universe, known to God for all ages, was drawn out over time.

While it is important to proceed, as the philosopher Mariusz Tabaczek points out, "with all caution and awareness of methodological differences between natural science and metaphysics,"[34] there seems to be a resonance between the language used by Thomistic philosophers and the language used by many modern evolutionary biologists. Both in the emergence of biological forms from the potential inherent in the physics and chemistry of the universe and in the emergence of more complex biological forms from the potential inherent in less complex forms, modern biology and Thomistic philosophy seem poised for a mutually enriching engagement.

HOW DO EVOLUTIONARY TRANSITIONS WORK? SUBSTANTIAL FORM VS. ACCIDENTAL FORM

While there does seem to be some general consilience between evolutionary thought and Thomistic metaphysics, when it comes down to the details, the successful integration of the two is challenging. One particular stumbling block is that for Aquinas, the substantial form of an organism was not something that was subject to change: it described that which made a frog a frog or a squirrel a squirrel. What could change for Aquinas were the accidental properties, such as the skin tone of the frog or the length of the squirrel's tail. Despite the fact that these accidental

34. Tabaczek, "What Do God and Creatures Really Do in Evolutionary Change?"

changes could occur from generation to generation, the substantial form would remain unaltered.

Thomistic critics of evolution, such as the Dominican Fr. Michael Chaberek, have argued that this distinction between variable accidental properties and an underlying stable substantial form negates the very possibility of the evolution of new forms. Their argument is that the biological process of evolution involves the alterations of accidental properties, size, shape, metabolic rate, etc., while the production a new species (from a Thomistic perspective) requires the production of a new substantial form. One can change the accidental characteristics all one wants, but the underlying substantial form is independent of these and therefore cannot change as the result of the mere accumulation of accidental changes.[35]

There are a variety of problems with this argument, but the most fundamental problem is that it blurs the distinction—something cautioned about in the previous section—between scientific and philosophical explanations. It is important to remember that when scientists discuss evolution and the origin of species, they are investigating a different question than the one Aquinas addresses when he discusses substantial forms. Evolutionary biologists are examining the material properties of the organisms involved and the physical interactions that these organisms have with their environment. They attempt to quantify and measure the physical changes that have occurred to individual organisms within a population such that they become morphologically distinct and/or reproductively isolated from other members of the population.

On the other hand, Aquinas employs the concept of substantial form to describe the specificity of a certain type of being.

35. For the most comprehensive response to Chabarek's claims, see Fr. Mariusz Tabaczek's *Theistic Evolution: A Contemporary Aristotelian-Thomistic Perspective* (Cambridge University Press, 2023) and "A Contemporary Aristotelian-Thomistic Perspective on the Evolutionary View of Reality and Theistic Evolution," *Religions* 15, no. 5 (2024): 524, https://doi.org/10.3390/rel15050524.

Aquinas is asking a nonscientific question, yet one that is more fundamental. He is concerned with why this specific instance of matter exists in the form of a rose bush or that specific instance exists in the form of a zebra. It is not a scientific principle he is after, but rather a metaphysical one.

When biologists discuss the transformation of species, the metaphysics regarding substantial form does not even enter the discussion. In fact, it is not something that biologists (acting as biologists) investigate because a substantial change is not a physical thing that is amenable to the investigative tools of the natural sciences. The substantial form of a duck is not something one can measure, weigh, or quantify. It is something one can comprehend—one knows one is looking at a duck and not a frog—but it is not something that can be quantified in any meaningful way. One cannot compare the substantial forms of a duck and a frog in the manner one can compare the physical pressure exerted by the heart chambers in a frog and a duck. The former deals with a metaphysical concept, the latter with a physical concept.

Despite these important differences, the two concepts are not unrelated, as substantial forms only make themselves known to us in physical, concrete entities. We would have no way of comprehending the form of a duck if no one had ever seen and observed an actual physical duck. Despite this relationship, the substantial form, though it is made manifest in the physical world, cannot be fully reduced to the physical and concrete.

Yet at the same time, physical matter has a critical relationship to the substantial form. In fact, Aquinas acknowledged that changes to physical matter can be *associated* with substantial changes. He is quite clear on this in his discussion of the burning of a log or the death of a human. In both of these cases, the material change is associated with one substantial form ceasing to be and another, which existed in potency, emerging (in the case of the burning of a log, the wood and the ash respectively).

An important distinction must be made here. While a scientist can give a mechanistic description of how wood is converted into ash, there is no mechanistic description possible of how the substantial form of wood can be "converted" to the substantial form of ash and smoke. In fact, for Aquinas, they don't actually convert or transform at all. Instead, in the case of the burning log, one substantial form merely ceases to exist while another is educed from the new arrangement of matter in the smoke and ash. One substantial form is *not* transforming into the other.

While he is clear that one substance, under certain physical conditions, can participate in the emergence of a different substantial form that it contains in potency, he does not discuss how this might occur in the emergence of new species. That Aquinas does not elaborate on this possibility is not surprising given that there was little evidence available to him that such organismal transitions might indeed occur. However, the fact that Aquinas did not attempt to develop his metaphysics to accommodate evolutionary transitions does not mean that evolutionary transitions cannot be reconciled with Aquinas' philosophy.

To investigate the ability of Aquinas' metaphysics to handle such a transition, it is worth looking at a specific example: a novel hybrid plant species. A novel hybrid plant involves a combination of genetic and molecular material that it inherits from its two parents, each of which is a member of a distinct species. Given the novel material composition of the hybrid, it can look and interact with its environment in a manner that is different than that of its parents. In fact, many successful hybrid species thrive in an environmental niche to which neither parent is particularly well adapted. Thus, these hybrids are often considered by biologists to be a distinct species. Biologically speaking, one can argue that the novel material composition of the hybrid zygote gave rise to an entity that has properties distinct from its parents and should be considered a separate species.

How then does this translate into philosophical language? Does the hybrid have a distinct substantial form compared to the parents? The scientist, as a scientist, is unable to resolve this question. He or she merely describes the new species' biological form (again, something wholly distinct from Aquinas' metaphysical concept of substantial form) and attempts to determine whether or not it should be considered a distinct species based upon the criteria biologists use to make this judgment.[36] However, there is nothing precluding a philosopher from arguing that the novel material composition of the hybrid zygote predisposes prime matter such that a new form is educed or made manifest. In fact, such a claim is not inconsistent with Thomas' understanding of nature, as Tabaczek points out:

> The process of evolution can be explained as a series of existential realizations of [substantial] forms in nature carried out through the process in which primary matter becomes properly disposed to be informed by new substantial forms. Species can thus be said to gradually change (evolve) in time [or rapidly in the case of hybrids]; but not without qualification. What needs to be clarified is that what actually change are accidental traits and properties of concrete organisms, which bring, in turn, an alteration of the disposition of primary matter, preparing it to receive the form of a new species. Therefore, strictly speaking, what the complex nexus of evolutionary processes brings about, from the Aristotelian [and Thomistic] point of view, is an existential realization of species as forms, educed from the potency of prime matter.[37]

36. It should be noted that biologists have a difficult time in determining exactly what criteria should be used to distinguish two populations as separate species. Whether it is reproductive isolation, morphological diversity, or genetic diversity, there is no clear-cut criterion that can be applied in an unequivocal fashion to make these determinations. This does not imply that species do not exist, only that in practice they may be difficult to unambiguously delineate. Nature is often more mysterious than our classification schemes.

37. Mariusz Tabaczek, "Thomistic Response to the Theory of Evolution: Aquinas on Natural Selection and the Perfection of the Universe," *Theology and Science* 13, no. 3 (2015): 328.

Consistent with the Aristotelean-Thomistic tradition, these forms can even be seen as unchanging and immutable because the substantial forms themselves are not being altered. Rather, in the case of the hybrid example above, the substantial forms of the parents still exist in the parents, but another type of form is educed from the potentiality of prime matter and is made manifest in the hybrid offspring.

To argue that evolution cannot produce novel species because, philosophically speaking, sequential accidental changes cannot add up to a substantial change is to conflate two different modes of explanation. When an evolutionary biologist posits that specific physical changes in organisms over time have led to the production of new species, he or she is not positing anything regarding changes to what Aquinas calls the substantial form. That is an explanation on an entirely different level. He or she is merely positing how a "complex nexus of evolutionary processes," processes that can be studied by scientists, could give rise to a new physical entity—namely, a new biological species.

While it is important to make distinctions between biological and philosophical explanations, this does not mean that they are unrelated. (They must be related to some extent if they both hope to refer to the same unified reality.) To integrate these two different levels of explanation, one merely needs to posit that the biological/physical novelty that occurs within a new organism (the novel hybrid) can be associated with the emergence of a new substantial form. Rather than positing that such and such a physical change *caused* the new substantial form, all one needs to posit is that the physical changes and conditions (the complex nexus of evolutionary processes) can predispose prime matter to receive the novel substantial form. The biology is not producing the substantial form; it is simply allowing the form to be made manifest through whatever physical changes have occurred. In addition, such a philosophical understanding of the emergence of the new substantial form does not contradict the biological explanation for the production of the hybrid species—namely, that certain physical changes have occurred such that a new biological species arises. Both the biological and the philosophical explanations work on their own terms and at their own proper levels.[38]

The relationship between the combination of matter and its associated substantial form is not an esoteric question but one that has specific relevance to modern molecular biology. Given our current ability to genetically engineer organisms, modern molecular biology raises a thicket of questions that illustrates the complex relationship between physical matter and substantial form. Imagine a biologist who successfully removes the DNA from a chimpanzee oocyte and replaces it with a full complement of human DNA. Would this entity have the substantial form of a human? Would it have the substantial form of a chimp? Would it be a new biological species with a novel substantial form? The answers to these are not obvious from either a scientific perspective or a philosophical perspective. Merely having human DNA does not necessarily mean the chimp oocyte will develop along the human trajectory with human capacities. Likewise, the fact that it is a chimp oocyte with chimp proteins does not ensure it will develop along the chimp trajectory given the inclusion of the human chromosome set.

Despite these uncertainties, one can be confident that a novel material arrangement would be created here, an arrangement that, philosophically speaking, could be disposed to make manifest an entirely new form. Yet, just as with the origin of a new species during the evolutionary process, this new substantial form would not come into being via the material transformation of one substantial form into another. The new substantial form, if

38. Determining what an "accidental change" is scientifically is problematic. Is a major chromosomal change in a zygote an "accidental change" or is it predisposing matter to a new substantial form? Does it facilitate a substantial change? At some point, if you disrupt the zygote enough, it is no longer the type of substance it once was. Exactly where one crosses this line is not clear.

indeed this entity would be informed by a new substantial form, would simply be made manifest through the novel arrangement of the underlying matter. In this manner, just as Aquinas believed that the specific arrangement of matter in the slime could, through the power of the sun, educe the substantial form of a maggot or a fly, the specific material arrangements of a novel hybrid zygote or a hypothetical human-chimp chimeric embryo could educe the substantial forms of new species.

AQUINAS AND EVOLUTION

In his *Dialogue Concerning Two Chief World Systems*, Galileo called to task the Aristotelians of his day for refusing to engage properly with the evidence for the corruptibility of the heavens. The observations of craters on the moon and sunspots on the sun indicated that corruptibility and change occurred not only on the Earth but in the heavenly sphere as well. Yet, despite this evidence, many still clung to Aristotle's position regarding the incorruptibility of the heavens. Ironically, clinging to such a position in the face of sensory evidence to the contrary put these Aristotelians in a distinctly anti-Aristotelian position. Galileo masterfully pointed out this inconsistency in the *Dialogue*:

> Then of the two propositions, both of them Aristotelian doctrines, the second—which says it is necessary to prefer the senses over arguments—is a more solid and definite doctrine than the other, which holds the heavens to be inalterable. Therefore it is better Aristotelian philosophy to say, "Heaven is alterable because my senses tell me so," than to say, "Heaven is inalterable because Aristotle was so persuaded by reasoning." Add to this that we possess a better basis for reasoning about celestial things than Aristotle did. He admitted such perceptions to be very difficult for him by reason of the distance from his senses, and conceded that one whose senses could better

> represent them would be able to philosophize about them with more certainty. Now we, thanks to the telescope, have brought the heavens thirty or forty times closer to us than they were to Aristotle, so that we can discern many things in them that he could not see; among other things these sunspots, which were absolutely invisible to him. Therefore we can treat of the heavens and the sun more confidently than Aristotle could.[39]

The Aristotelians of Galileo's day were so committed to Aristotle's philosophical reasoning on the heavens that they refused to accept any new knowledge that would require the rethinking of Aristotle's argument—despite Aristotle's own insistence on the primary role of sensory evidence.

In a similar fashion, Aquinas' statements regarding the fixity of the forms of species and the need for direct creation of certain species by God should not supersede any new evidence for the transformation of species, evidence that Aquinas did not have at his disposal. It seems quite likely that Aquinas himself might have taken this new sensory evidence at face value and attempted to integrate it into his philosophical system, particularly because Aquinas, like Aristotle before him, believed that intellectual knowledge flowed in some sense from sensory cognition.

According to Aquinas, "The object of every sensitive power is a form as existing in corporeal matter."[40] From this perception of the material world, the human intellect could proceed to understanding: "Our intellect understands material things by abstracting from the phantasms; and through material things thus considered we acquire some knowledge of immaterial things."[41]

Given his position, there is no reason to suspect that Aquinas would simply object on philosophical grounds and ignore the

39. Galileo Galilei, "The First Day," in *Dialogue Concerning the Two Chief World Systems*, 2nd ed., trans. Stillman Drake (Berkeley: University of California Press, 1967), 55–56.

40. *ST* 1.85.1 co.

41. *ST* 1.85.1 co.

sensory evidence if the evidence in his day had suggested that biological species originated via evolutionary processes. In fact, his acknowledgment that new species such as hybrids could arise in certain cases flowed directly from the sensory evidence he did have at hand. Given that the origin of hybrid organisms had indeed been observed, Aquinas successfully integrated this observation into his metaphysics. He argued that new hybrid species "existed beforehand in various active powers"[42] of other created entities. In other words, the potential for them was inherent in other entities that were created during the Genesis creation narrative. While this is far from an evolutionary view of the world, the reasoning seems easily applicable to the evolutionary origin of other species. Any species could have "pre-existed in certain active powers" of other created entities. Such a position, in fact, resonates with the scientific evidence that suggests that the potentialities for elements, molecules, cells, and complex life forms have been embedded within the very fabric of the created world from the beginning of time.

The importance that sensory observation had in shaping Aquinas' understanding of the generation of forms is not limited to the case of hybrids. Sensory evidence also drove Aquinas' philosophical position on putrefaction. Based on the observations of the time, it appeared that simple animals could be generated spontaneously from the slime via the heating power of the sun. However, no one had observed more complex organisms emerging from the slime. As a result, Aquinas argued that "the power of heavenly bodies suffices for the production of some imperfect animals from properly disposed matter: for it is clear that more conditions are required to produce a perfect than an imperfect thing."[43] Basically, he argued that animals such as lions and tigers were too "perfect" to be generated solely by the power of the sun, since they never emerged "spontaneously" out of slime in the

42. ST 1.73.1 ad. 3.

43. *ST* 1.91.2 ad. 3.

manner that maggots did. Thus, these higher forms required the focused heating of the semen from a similar animal to act as an instrumental cause through which the principal agent, the sun, could bring about the generation of the animal. In both cases, though, it was through created causes that the new organisms arose, causes that Aquinas believed one could observe (even if he interpreted those observations incorrectly).

Aquinas, like Aristotle before him, was determined to ground the origin of new forms within the material world, as opposed to Plato, who grounded them in an immaterial world of forms inaccessible to our senses.[44] It is ironic then that some Thomistic critics of evolution have sought to demonstrate that created causes cannot explain the emergence of new species, arguing rather that such species must directly be created by God. As Carl points out, "Perhaps this is the metaphysical conclusion at which one needs to arrive; but if so, we will have left Aristotelian or Thomistic metaphysics behind on one of its central conclusions about natural substances."[45] What we will have left behind is the Aristotelian-Thomistic notion that causes within the natural world are sufficient to explain the generation of living organisms. Rather than leaving this behind, it seems more apt to, as Carl states, "be optimistic with Thomas that it pertains to the goodness of the divine to endow creation with the power to act as a genuine cause of anything moved or generated."[46] If we take this view, there seems to be nothing within Aquinas' philosophy that would make it, in principle, incompatible with an evolutionary understanding of the biosphere. On the contrary, there may indeed exist a deep synergy between the two.

44. Aquinas also connected the forms to the mind of God as explained above in relation to his concept of God as the exemplar cause of forms. However, their actual manifestation in the world was attributed to natural processes, as described previously.

45. Carl, "Thomas Aquinas on the Proportionate Causes of Living Species," 246.

46. Carl, 246.

8

Human Origins

From the Dust of the Earth

The biological evolution has transcended itself in the human "revolution." A new level or dimension has been reached.[1]

—Theodosius Dobzhansky

Given the profound theological implications evolutionary theory had regarding the nature of man, it's not surprising that most of the early opposition regarding evolution from within Catholic circles centered around the connection between man and other primates. In fact, the earliest reference to evolution by the Church, a decree from the Provincial Council of Cologne in 1860 (which was put forth only a year after Darwin published his *Origin of Species*), denounced the "spontaneous transformation" of a primate into a human. As discussed in chapter 3, this concern regarding human origins did not soon dissipate. Even Pope Pius XII, the author of the 1950 encyclical *Humani Generis*, was somewhat skeptical of an evolutionary origin of man, cautioning against those who "act as if the origin of the human body from pre-existing and living matter were already completely certain and proved by the facts."[2]

Darwin himself understood he would encounter significant resistance to the inclusion of man in the evolutionary process. In fact, he decided not to explicitly broach the subject in the *Origin of Species* (1859) apart from a brief mention that through his theory, "much light will be thrown on the origin of man and

1. Theodosius Dobzhansky, *The Biology of Ultimate Concern* (New York: New American Library, 1967), 58.

2. Pius XII, *Humani Generis* 36, encyclical letter, August 12, 1950, vatican.va.

his history."[3] While this did not prevent others from discussing the implications his theory had for human origins, Darwin saved his thoughts on the matter for his later publication, *The Descent of Man* (1871). It was in *Descent* that Darwin laid bare his ideas regarding the evolutionary origin of man, including his explanation for how man's moral sense and man's impressive cognitive abilities could have evolved from other primates.

While both *Origin* and *Descent* argue that natural selection can explain the biological origin of species—a position *not* at odds with Catholic teaching—it is in *Descent* that Darwin argues that all of man's mental abilities, including his moral sense, are nothing more than the product of an evolutionary process. Darwin saw "no fundamental difference between man and the higher mammals in their mental faculties,"[4] despite the fact that the other primates completely lack things like syntactical language, religion, art, literature, commerce, agriculture, science, and politics. Despite these significant differences, Darwin argued that man's ability to self-reflect, his ability to reason about abstract concepts like love and justice, and his innate moral compass evolved gradually from other animals via naturalistic means.

Darwin's view of human morality is particularly instructive on this point, as he believed "morality" could be fully explained as the natural outgrowth of the social instincts found in other advanced primates. For example, in the small prehuman hominin populations in the African savannah, survival outside the social group would have been difficult. As a result, individuals would have been under strong pressure to follow certain social instincts and maintain peace and stability within the group. Over time, as primitive hominins evolved, they would regard following through on these social instincts as "moral" and the succumbing to baser instincts—like stealing from the group's food supply—as

3. Charles Darwin, *The Origin of Species* (New York: Mentor, 1958), 458.

4. Charles Darwin, *The Descent of Man, and Selection in Relation to Sex* (London: Penguin Classics, 2004), chapter 3.

"immoral." In this case, "morality" is reduced to nothing more than our group survival instincts triggering the emotions of guilt and shame in a particular setting in East Africa.

Based upon this type of reasoning, those acts we deem to be moral would vary depending upon the type of social instincts that become evolutionarily important for survival. If the social instinct to kill the disabled members of the group became beneficial for the survival of the group, then such acts would be deemed "moral" and we would evolve the proper "moral" emotions around such behavior.

While Darwin's view on the origins of morality may seem to put his theory at odds with Church teaching, one has to be careful to separate Darwin's views on the subject of human morality from what evolutionary science can actually demonstrate. While there is certainly some truth to the notion that animal emotions and instincts have had a role in shaping our moral sensibilities, there is good reason to believe that naturalistic explanations cannot fully explain our moral nature. Darwin, however, wanted man to be explained entirely by the evolutionary process. He wanted man to be nothing more than a material being fully explicable by the scientific method.

This view, typically called Darwin*ism*, is held by many evolutionary scientists, but it is ultimately not a scientific position. Darwin's theory (or any other evolutionary theory), as a *scientific* theory, solely studies the material processes of evolution and how these effect the emergence of the physical properties of organisms. As a scientific theory, it is incapable of demonstrating the existence or nonexistence of the immaterial aspects of man. Thus, when someone like the biologist Richard Dawkins makes the claim that "the sudden injection of an immortal soul in the time-line is an anti-evolutionary intrusion into the domain of science," he misses the mark.[5] The proposed existence of a human

5. Richard Dawkins, "You Can't Have It Both Ways: Irreconcilable Differences," in *Science and Religion: Are They Compatible?* ed. Paul Kurtz (Buffalo, NY: Prometheus Books, 2003), 208.

soul is not anti-evolutionary; rather, it is something that goes beyond evolution. Evolution, as a scientific theory, cannot determine whether such an immaterial thing exists or whether such an event occurs. This fact does not preclude people like Dawkins from speculating or arguing that there is no immaterial aspect to man or that man is reducible to matter. But it is important to recognize that they have crossed over into the domain of philosophy. If Dawkins wishes to refute the idea that man is a unity of body and soul, he must develop a sophisticated philosophical argument to bolster his case rather than just stating that "science" doesn't allow such things. What doesn't allow such things is scientism (a philosophical position), not science.

It is here, at the junction of science and philosophy, that the big questions regarding existence are explored, and it is here that science and philosophy can help inform each other. Good philosophy can help clarify scientific conundrums, and good science can help ground philosophy. In fact, scientific data can be used to support philosophical arguments. For example, a materialist could claim that the physical activity in the brain can explain man's moral sense. He could argue that what we perceive as "morality" is merely the result of neurons firing in certain regions of the brain. He could study (scientifically) the electrical changes in the brain to support his argument. He could even perform scientific research to identify that there is activity in one specific region of the brain every time someone senses that an act is "immoral." Such scientific data may be consistent with his argument, but it doesn't prove it. In fact, one could never prove it scientifically because the argument he is making goes beyond the domain and competence of science. People like Dawkins are claiming that the physical changes in the brain *entirely* explain morality. However, it is quite possible that while the physical changes in the brain may be necessary for moral reasoning, they are not sufficient. It is entirely possible that other immaterial aspects of the human person are needed for this. The fact of the matter is that there is no scientific experiment capable of distinguishing between these two positions. This is a key point. While scientific data can be used to support such arguments, it is not able to resolve them. One has to justify whichever position one takes using philosophical reasoning.

CATHOLIC ANTHROPOLOGY

While the science of human evolution is both fascinating and complex, it is important to realize that science cannot answer the question of what man is from a theological and philosophical perspective. How man's physical form has evolved from other hominin species is certainly relevant to an understanding of the human person, but it is not sufficient to explain the human person in his entirety. Likewise, the Genesis creation accounts of the origin of man are not meant to be taken as historical descriptions that explain how man's physical body emerged but as articulations of truths regarding the nature of man. As Ratzinger points out, Genesis "does not in fact explain how human persons come to be but rather what they are."[6] It is in explaining what humans are—namely, creatures made in the image and likeness of God—that the text of Genesis finds relevance to discussions regarding evolution.

The first truth that Genesis makes clear, again, is that man is created in the image and likeness of God. This means that human beings are not some meaningless by-products of an evolutionary process. God created us for a reason: to know him, to love him, and to be with him in heaven. The fact that we are created in the image and likeness of God should not be interpreted to mean that we look like God (or that God looks like us). Rather, this phrase indicates that we possess an inherent dignity in that we alone, of all material creatures, have the capacity to know and love him. Of

6. Joseph Ratzinger, *"In the Beginning . . .": A Catholic Understanding of the Story of Creation and the Fall* (Grand Rapids, MI: Eerdmans, 1995), 50.

all earthly creatures, only mankind has been called to love God freely as God has loved us. It is in this relationship with God that we reflect, however imperfectly, his image.

Such a view of man also indicates that we cannot be reduced to purely material beings. Instead, man is, as the *Catechism* states, "a being at once corporeal and spiritual."[7] The image in Genesis, in which God forms man from the dust and then animates him with the breath of life, is indicative of this reality; man is a unity of body and soul. In addition, the fact that the soul is an immaterial entity places it outside of the scope of the evolutionary process. Pope Pius XII made this exact point in *Humani Generis* when he indicated that while evolution *might* explain "the origin of the human body as coming from pre-existent and living matter,"[8] Catholics must still "hold that souls are immediately created by God."[9] That we are an ensouled body, more than any other aspect of our existence, is indicative of our being made in the image and likeness of God.

While the Church teaches that man is "a being at once corporeal and spiritual," Catholics do not take this position simply on faith. There are many distinctive features of man—his ability to engage in conceptual thought, his ability to abstract immaterial universal concepts, his perception of universal moral precepts, his sense of free will, and his rich internal conscious experience—that are difficult to explain satisfactorily within a purely materialistic framework. This is not the place to litigate these arguments—only to make clear that there are robust rational arguments to support the position that man is more than a material being. Thus, the Catholic position that man is a unity of body and soul is an eminently reasonable stance that is consistent with our everyday experience as free, rational, moral agents.

Genesis also has something to say about our original

7. *Catechism of the Catholic Church* 362.

8. *Humani Generis* 36.

9. *Humani Generis* 36.

relationship to God and creation. There are two key points here that are worth highlighting in the context of human evolution. The first is that man was given the freedom to choose to love God and neighbor. The second is that man was initially free from the bounds of sin and was "in friendship with his Creator and in harmony with himself and with the creation around him."[10] After the creation of mankind, the Genesis account reads, "God saw everything that he had made, and indeed, it was very good" (Gen. 1:31). At the outset, man was both free and sinless, but man abused his freedom and sin entered the world. This is reflected in the story of Adam and Eve in the garden, a symbolically rich description of what the *Catechism* refers to as "a primeval event, a deed that took place at the beginning of the history of man."[11]

In summary then, here are the four major points that Genesis reveals regarding man that are key to any discussion of the evolutionary origin of man:

- God created man in his own image and likeness.
- Man is a unity of body and soul. The soul is immaterial, immortal, and created directly by God. The soul did not evolve.
- Man was endowed with freedom to choose to love God and neighbor.
- Man was created in a state of harmony with God, free from the corruption of sin.

With these truths in mind, it is time to examine what modern science has uncovered regarding our evolutionary history.

10. *CCC* 374.
11. *CCC* 390.

THE HOMININ FOSSIL RECORD

The story of human origins that the fossil record reveals is complex, contentious, and subject to continuous revision as new fossil finds are unearthed.[12] For example, up until recently, the earliest known modern human fossils had been dated to about 200,000 years ago and had been found in Ethiopia in the east of Africa.[13] In 2017, though, a team of paleontologists dated a set of modern human fossil remains from Morocco to roughly 300,000 years of age.[14] This Moroccan fossil find (along with other recently identified human remains from South Africa dated to 260,000 years of age) has disrupted the consensus belief that the first modern humans evolved in East Africa and has significantly complicated the picture of early human evolution.

Such disagreements and shifting opinions are typical regarding the hominin fossil record, which spans the last six to seven million years and includes all the fossils that are more closely related to modern humans than to modern chimpanzees. As a result, any attempt to summarize the hominin fossil record is fraught with difficulties. Both the discovery of new fossil finds and the difficulties associated with unambiguously classifying the fossil remains that have already been uncovered make nearly any general summary of the record contentious. Despite these difficulties, it is worth attempting to paint a broad picture of the current state of the field.

The earliest hominin fossils date to about 6.5 million years ago and are found solely in Africa, predominately East Africa. The fossil record is particularly sparse during this early period (4.5

12. For an accessible overview of the recent hominin record, see Brian Handwerk, "An Evolutionary Timeline of Homo Sapiens," *Smithsonian Magazine*, February 2, 2021, https://www.smithsonianmag.com/science-nature/essential-timeline-understanding-evolution-homo-sapiens-180976807/.

13. Ian McDougall et al., "Stratigraphic Placement and Age of Modern Humans from Kibish, Ethiopia," *Nature* 433 (2005): 733–736, https://doi.org/10.1038/nature03258.

14. E. Callaway, "Oldest *Homo Sapiens* Fossil Claim Rewrites Our Species' History," *Nature* (2017), https://doi.org/10.1038/nature.2017.22114.

to 7 million years ago), so not much is known definitively about the hominin ancestors that existed during this time, although some seem to have skeletal features associated with a transition to bipedal locomotion, such as a forward shift in the location of the foramen magnum (the opening in the skull for the spinal cord). There are at least four different hominin species that have been uncovered from this period, but they are mostly incomplete and fragmentary finds.[15]

The hominin fossil picture changed dramatically around 4.5 million years ago as a host of fossil species belonging to the genus Australopithecus began to appear in the record. At least seven different species of Australopithecus are recognized, and like earlier hominin fossils, these have been found exclusively in Africa. The australopith fossils had brain sizes similar to those of modern chimps (400 ccs—about one-third of the size of modern humans) but appear to have been relatively adept at bipedal locomotion based on their skeletal qualities. The well-known fossil Lucy (*Australopithecus afarensis*) is a member of this genus and has been dated to about 3.2 million years ago. The australopiths persisted in the fossil record for nearly two million years and produced a variety of forms. Our genus Homo is thought to have evolved from one of these forms, although there is debate regarding which of the australopiths identified thus far gave rise to the genus Homo.[16]

The earliest member of our genus, *Homo habilis*, emerged in the fossil record around 2.3 million years ago. While *Homo habilis* has some distinctive facial and tooth features that distinguish it from the australopiths, its body and brain sizes overlap with them. As a result, there is some debate regarding into which

15. For an easy-to-read general overview, see D.F. Su, "The Earliest Hominins: Sahelanthropus, Orrorin, and Ardipithecus," *Nature Education Knowledge* 4, no. 4 (2013): 11, https://www.nature.com/scitable/knowledge/library/the-earliest-hominins-sahelanthropus-orrorin-and-ardipithecus-67648286/.

16. For an easy-to-read general overview, see C.V. Ward and A.S. Hammond, "Australopithecus and Kin," *Nature Education Knowledge* 7, no. 3 (2016): 1, https://www.nature.com/scitable/knowledge/library/australopithecus-and-kin-145077614/.

genus it should be classified.[17] A number of other members of the genus *Homo* emerged just after this time, all of which had more modern features than the australopiths, such as larger brain sizes, smaller jaws, and more modern leg and arm proportions. The most impressive of these is *Homo erectus*, which can be found in the fossil record from nearly two million years ago up until roughly 150,000 years ago. *Homo erectus* has the honor of being the first hominin species to have ventured out of Africa, and its remains have been found in locales ranging from southern Europe to Asia to Indonesia. *Homo erectus* was significantly larger than the australopiths and had a brain capacity (about 800 to 1000 ccs) closer to that found in modern humans (about 1300 ccs). Their wide temporal and geographical distribution was associated with a variety of different populations, with the *Homo erectus* in Africa having a slightly different set of traits as compared to those found in Indonesia. In fact, some have argued that the *Homo erectus* from Africa and western Asia should be classified as a separate species, *Homo ergaster*, thereby differentiating them from the *Homo erectus* fossils found in eastern Asia and Indonesia.[18]

Regardless of the outcome of this dispute, the *Homo erectus* variants in Africa are thought to have given rise to a new species known as *Homo heidelbergensis* roughly 600,000 years ago. Members of *Homo heidelbergensis*, also known as archaic *Homo sapiens*, were about 1.5 meters tall and had brain sizes at the lower end of the range seen with modern humans (about 1200 ccs). Fossils lumped into this grouping have been found in Africa, Europe, and Asia, mimicking the range of *Homo erectus*. Those that migrated to Europe and Eurasia are thought to have given rise to the Neanderthals and other related species, while modern

17. M. Collard and B.A. Wood, "Defining the Genus *Homo*," in *Handbook of Paleoanthropology*, 2nd ed., eds. W. Henke and I. Tattersall (Berlin: Springer, 2015), 2108–2144.

18. Susan C. Anton et al., "Morphological Variation in *Homo erectus* and the Origins of Developmental Plasticity," *Philosophical Transactions of the Royal Society B* 371 (2016), http://doi.org/10.1098/rstb.2015.0236.

Homo sapiens are thought to have emerged from the population of archaic *Homo sapiens* that remained in Africa.[19]

As described above, the oldest anatomically modern *Homo sapiens* fossils have been found in North Africa and date to 300,000 years of age. Despite having a modern brain size (about 1300 ccs), these fossils display an elongated and flattened brain shape as compared to modern humans. This has led to some disagreement over whether they should be classified as modern *Homo sapiens* or a distinct species. There is, however, general agreement that a more recent fossil find from East Africa, which dates to about 200,000 years ago, represents modern *Homo sapiens*.

While modern *Homo sapiens* emerged in Africa between 200,000 to 300,000 years ago, members of our species soon left Africa and spread to every continent besides Antarctica. Modern *Homo sapiens* are thought to have first successfully ventured out of Africa roughly 120,000 years ago.[20] There is some evidence that earlier migrants entered the Middle East around 200,000 years ago, but they do not appear to have gained a foothold in the region. Later migrants that left Africa around 120,000 years ago were more successful in their ventures and eventually spread to Asia by at least 80,000 years ago. These early human migrants did not immediately displace the other hominins they encountered in Asia or the Levant, though there is evidence that they interbred with some of them.[21]

While there was likely movement of modern humans in and out of Africa throughout prehistoric times, a second major dispersal seems to have left Africa around 50,000 to 60,000 years ago. This later wave brought humans into Europe, Australia, and

19. Bridget Alex, "*Homo heidelbergensis*: The Answer to a Mysterious Period in Human History?" *Discover Magazine*, September 16, 2009, https://www.discovermagazine.com/planet-earth/homo-heidelbergensis-the-answer-to-a-mysterious-period-in-human-history.

20. S. Tucci and J. Akey, "A Map of Human Wanderlust," *Nature* 538 (2016): 179–180, https://doi.org/10.1038/nature19472.

21. R. Nielsen et al., "Tracing the Peopling of the World through Genomics," *Nature* 541 (2017): 302–310, https://doi.org/10.1038/nature21347.

the New World, and eventually contributed to the displacement of all the other hominin species throughout the globe.[22]

THEOLOGICAL VS. BIOLOGICAL HUMANS

While the fossil record indicates that individuals with skeletal characteristics similar to those of modern humans may have existed for 300,000 years, it's not necessarily clear that these individuals were human in the same sense as modern-day humans. To understand this point, it is useful to make a distinction between individuals that are considered *theologically human* and individuals classified as *biologically human*.[23] At first glance, this may not seem like a useful distinction given that every individual living today who is classified biologically as a *Homo sapiens* is considered theologically human. However, these two classifications may not overlap perfectly in the fossil record.

To illustrate this issue, it's necessary to define these terms clearly. For starters, to be considered a theological human, one must be the type of creature for which it is proper to be informed by a rational human soul. Such creatures are made in the image and likeness of God and are, at least in principle, capable of conceptual thought, syntactical language, and worship of the Creator. While theological humans would need to have the type of body that was fitting to be informed by a rational soul, what the exact parameters for this body might be is not clear. Would the Moroccan fossils from 300,000 years ago have the type of body that was fitting to be endowed with a rational soul, or would the difference in their braincase exclude them from being theologically human? The short answer is no one knows.

A similar issue exists with the term biological humans. For a fossil to be considered biologically human (to be classified as

22. Neilsen et al., "Tracing the Peopling of the World."

23. This distinction is made by Kenneth Kemp in his article "Science, Theology, and Monogenesis," *American Catholic Philosophical Quarterly* 85, no. 2 (2011).

Homo sapiens), scientists must conclude that it has a skeletal structure that is similar enough to modern humans. This is a judgment call. How does one know for certain when the anatomical structures uncovered in a fossil are similar enough to the range seen in modern humans such that it should be classified *biologically* as a member of the species *Homo sapiens*? The "human" fossils dated to 300,000 look similar to modern humans, but there are some distinct differences, particularly in the shape of the braincase. Given these differences, it is unclear whether they should be classified biologically as modern *Homo sapiens* or rather as a separate species or subspecies. Not surprisingly, difficulties and disputes regarding classification are quite common in paleontology. This is particularly the case with the hominin fossils, given that 1) most fossil finds are fragmentary (they only consist of partial skeletal remains) and 2) fossil finds from the same species show significant variation depending upon the locale at which they are unearthed. (For example, *Homo erectus* fossils from Asia are distinct from those found in Africa.)

While there are obvious difficulties in determining whether specific fossils should be classified as biologically human, these difficulties pale in comparison to the difficulties in determining if a fossil should be classified as theologically human. Obviously, souls don't fossilize, and neither do key traits associated with theological humans, such as rational thought and language. As a result, it is always possible that fossils that look anatomically similar to modern humans and that scientists might classify as biologically human (for example, the Moroccan fossils) may have been different enough that they were not the type of creature fitting to be informed by a human soul (a theological human). From a theological perspective, they would be considered "pre-human," as they would presumably lack certain biological features that would make a theological human possible. While resolving this question in an unambiguous fashion is likely impossible given the inherent fragmentary nature of the fossil and archaeological records, it is possible to make some solid conjectures regarding the emergence of theological humans based upon the evidence.

WHO WERE THE FIRST THEOLOGICAL HUMANS?

While the determination of when theological humans first appeared on Earth is fraught with obstacles, the archaeological record can give some clues regarding an upper bound for when they emerged. There are certain artifacts we would only expect to be associated with beings capable of rational thought and syntactical language. Whenever these are found in the archaeological record, it is reasonable to assume that theological humans were present to create them. Such artifacts include sophisticated tool-making techniques, figurative artwork, therianthropic imagery, musical instruments, decorative jewelry, and purposeful burial practices.

Prior to 100,000 years ago, artifacts of a clearly symbolic nature are generally lacking in the archaeological record. Around this time, though, artifacts suggesting symbolic thought begin to make a sporadic appearance at sites in Africa and the Middle East. The artifacts that have been found include jewelry, engravings, hafting techniques (the attachment of blades to handles), and pigment use.[24] These examples tend to be diffusely spread throughout various sites, although a number of them, like engraved ochre and composite tools, have been found at the Blombos cave in South Africa.[25] Given the Blombos cave find and the discovery of similar artifacts at additional South African locales, some theorists have posited that modern human behaviors had their birth in South Africa before eventually spreading throughout the continent and beyond. However, evidence for

24. For an easy-to-read overview, see S. Wurz, "The Transition to Modern Behavior," *Nature Education Knowledge* 3, no. 10 (2012): 15, https://www.nature.com/scitable/knowledge/library/the-transition-to-modern-behavior-86614339/.

25. Kristen Tylen et al., "The Evolution of Early Symbolic Behavior in Homo sapiens," *PNAS* 117, no. 9 (2020): 4578–4584.

decorative jewelry, burial practices, and the use of pigments from roughly 100,000 years ago has been found in the Middle East, calling into question the idea that human culture emerged in one specific locale. The current evidence seems to support the notion that modern human behaviors appeared in the fossil record in a sporadic and mosaic fashion over a wide geographical area rather than a single locale from which all modern culture emerged.[26]

This sporadic emergence gave way to a major increase in human cultural artifacts around 40,000 to 50,000 years ago. Around this time, cave paintings, three-dimensional fertility carvings, musical instruments, and complex weapons began to appear at numerous European sites as well as in Indonesia.[27] In addition, findings from an African site dated to 44,000 years ago revealed jewelry (ostrich eggshell beads), bone-carved arrowhead points, and a lump of beeswax mixed with a poisonous resin that was wrapped in vegetable twine (most likely a component of a complex hunting instrument).[28]

While it is difficult to reconstruct their entire behavioral repertoire, it seems certain, given the associated artifacts, that the biological humans living around 50,000 years ago were also theologically human. In addition, modern humans are known to have reached Australia at around this time. While the sea levels were lower during this period, arriving in Australia from southeast Asia still would have involved an open sea voyage across a deep channel. It is hard to imagine how this could have been accomplished without a high degree of communication, planning, and

26. N.J. Conard, "A Critical View of the Evidence for a Southern African Origin of Behavioural Modernity," *South African Archaeological Society Goodwin Series* 10 (2008): 175–179; A. Meneganzin and A. Currie, "Behavioural Modernity, Investigative Disintegration & Rubicon Expectation," *Synthese* 200, no. 47 (2022), https://doi.org/10.1007/s11229-022-03491-7.

27. M. Aubert et al., "Earliest Hunting Scene in Prehistoric Art," *Nature* 576 (2019): 442–445, https://doi.org/10.1038/s41586-019-1806-y.

28. F. d'Errico et al., "Early Evidence of San Material Culture Represented by Organic Artifacts from Border Cave, South Africa," *Proceedings of the National Academy of Sciences of the United States of America* 109, no. 33 (2012): 13214–13219, https://doi.org/10.1073/pnas.1204213109.

foresight on the part of the voyagers, all skills associated with conceptual, rational thought.

This does not mean that the biological humans that existed earlier than this time were not theologically human. Given that they had a brain capacity very similar to modern humans, coupled with the sporadic evidence of primitive human cultural activity before 50,000 years ago, their being theologically human remains a distinct possibility. These humans may have had the mental and spiritual characteristics of modern humans but merely lacked the cultural knowledge of their descendants. Cultural knowledge—how to hunt specific prey, how to make sophisticated arrows, how to make music, how to carve and paint—is likely to have been assembled slowly, particularly in these small isolated human populations that lacked a written language.

While it remains possible that the fossils classified as biologically human from 200,000 to 300,000 years ago were also theologically human, identifying an exact timepoint for the emergence of theological humans is virtually impossible. It is always possible that these fossils, which look to be anatomically similar to modern humans, were in fact distinct enough such that they were theologically prehuman and not yet properly constituted to be informed by a rational soul. Further evidence from the paleoanthropological record may shed light on this question, but it is unlikely to fully resolve it.

There is one other interesting aspect of the archaeological record that impacts the question of what it means to be theologically human. This is the fact that there are other species such as *Homo neanderthalensis* that also seem to be associated with complex cultural artifacts and practices. In addition, there is good evidence that Neanderthals interbred with members of our own species. This raises interesting questions regarding what exactly it means to be human, both from a biological and a theological perspective.

HUMANS' INTERACTIONS WITH NEANDERTHALS AND OTHER ARCHAIC HUMANS

When modern humans left Africa, they encountered a number of other hominin species that had already established themselves in both Europe and Asia. How these encounters played out is not exactly clear, other than the obvious fact that these hominin populations were replaced eventually by the modern human migrants. While they encountered a variety of hominin species, it is the encounter between modern humans and Neanderthals that has generated the most attention.

The Neanderthals are believed to have descended from a group of hominins closely related to *Homo heidelbergensis*, which left Africa around a half a million years ago. The Neanderthals established themselves in the Levant, Eurasia, and eventually Europe; some even argue they made it to east Asia. The Neanderthals who settled in Europe appear to have had the continent to themselves until modern humans arrived sometime around 50,000 years ago.[29] Compared to modern humans, Neanderthals had more robust bodies—larger joints and a broader rib cage—but they had a brain capacity (1300 ccs) on parallel with ours. This has led some researchers to suggest that they had the ability for symbolic thought, much like modern humans.

Evidence for this includes primitive cave paintings from the La Pasiega cave in Spain that date to around 64,000 years ago, a date roughly 15,000 years before humans made their way into Europe.[30] There are also a variety of artifacts, including animal teeth with grooves and piercings (possibly to be worn as jewelry), that have been found in the Grotte du Renne cave in France. These artifacts were unearthed in a layer associated with Neanderthal

29. Ludovic Slimak et al., "Modern Human Incursion into Neanderthal Territories 54,000 Years Ago at Mandrin, France," *Science Advances* 8 (2022), https://doi.org/10.1126/sciadv.abj9496.

30. D.L. Hoffmann et al., "U-Th Dating of Carbonate Crusts Reveals Neandertal Origin of Iberian Cave Art," *Science* 359 (2018): 912–915, https://doi.org/10.1126/science.aap7778.

remains that dates to nearly 50,000 years ago.[31] In addition, there is evidence from the La Chapelle site in France that dates to the same time period that Neanderthals buried their dead.[32] A set of Neanderthal bones from this site were found buried in a depression that protected them from disturbance or alteration, unlike the unprotected animal bones found at the same site. In addition, the Neanderthal fossils from the site displayed signs of back and hip problems, suggesting that Neanderthals cared for their infirmed. There is even evidence for Neanderthal burial dating back to 70,000 years ago in the Middle East.[33]

While there is a body of evidence that can be used to support the notion that Neanderthals exhibited symbolic and conceptual thought, the evidence is fragmentary and is open to other interpretations. It is possible that the artifacts found at Neanderthal sites were acquired from encounters with humans rather than being fashioned by Neanderthals. Even in the cases suggesting Neanderthal burial practices, there are other alternative explanations. Given that there is evidence that 1) modern humans were burying their dead at least 100,000 years ago and 2) humans and Neanderthals likely interacted in the Middle East during this time period, it is possible that this behavior of Neanderthals was a form of imitation. Or it could have been a practice purely aimed at avoiding the odors of a decomposing corpse and therefore had no symbolic importance.

While there is some evidence for primitive art and jewelry as well as ritual burial at a few Neanderthal sites, what is absent is the range of complex, highly symbolic artifacts seen in the human record. As a result, the question of whether they had the ability for symbolic thought or whether these artifacts and behaviors

31. Frido Welker et al., "Palaeoproteomic Evidence Identifies Archaic Hominins Associated with the Chatelperronian at the Grotte du Renne," *PNAS* 113, no. 30 (2016), https://doi.org/10.1073/pnas.1605834113.

32. William Rendu et al., "Evidence Supporting an Intentional Neandertal Burial at La Chapelle-aux-Saints," *PNAS* 111, no. 1 (2014): 81–86.

33. E. Pomeroy et al., "New Neanderthal Remains Associated with the 'Flower Burial' at Shanidar Cave," *Antiquity* 94, no. 373 (2020): 11–26, https://doi.org/10.15184/aqy.2019.207.

were the result of copying human activities is open to debate. The fact that they seem to arise right around the time humans came onto the scene, particularly in Europe, is consistent with this possibility.

While it is not clear how much culture was transmitted between humans and Neanderthals, DNA sequencing data from both groups has revealed that genetic information was transmitted. In fact, roughly one to three percent of the DNA of non-African modern humans is of Neanderthal origin. It appears that the modern humans who left Africa mated with Neanderthals after they ventured out.[34] While the majority of the Neanderthal genes that entered the human gene pool due to these mating events were not maintained, a small portion have remained to the present day. The genes that were retained are thought to have given these early human migrants some selective benefits in their new environments outside of Africa. As more Neanderthal and ancient human genomes have been sequenced, more evidence for this type of mating has emerged. For example, genetic material isolated from 40,000-year-old human bones discovered in Romania contain about six to nine percent Neanderthal DNA, suggesting that this individual was only four generations removed from a successful human-Neanderthal mating.[35]

What is one to make of this from both a biological and a theological perspective? Based on the available evidence, it seems clear that humans and Neanderthals did interbreed and that at least some of their offspring had the ability to reproduce. Given that one scientific definition of a species is a group of individuals who interbreed and produce fertile offspring, should humans and Neanderthals be considered the same species biologically?

Merely having the ability to produce fertile offspring, though, is not always enough to be classified as a single species.

34. J.C. Teixeira and A. Cooper, "Using Hominin Introgressions to Trace Modern Human Dispersals," *PNAS* 116, no. 31 (2019): 15327–15332.

35. S. Mallick et al., "An Early Modern Human from Romania with a Recent Neanderthal Ancestor," *Nature* 524 (2015): 216–219, https://doi.org/10.1038/nature14558.

For example, lions and tigers can interbreed (with the female offspring in some cases being fertile), yet they are classified as distinct species. It appears that the situation with humans and Neanderthals is more analogous to the situation with lions and tigers in that they could interbreed in some cases but for the most part did not, instead remaining relatively genetically isolated from each other. In fact, despite the evidence for interbreeding, humans and Neanderthals maintained distinct morphologies in the fossil record despite coming into contact for tens of thousands of years.

Regardless, the data still suggests that modern theological humans did interbreed with another species, one that likely lacked a rational soul. Of course, it is possible that the Neanderthals were also theologically human, or the type of entity fitting to receive a rational soul. However, this seems a stretch given the current archaeological evidence as well as the fact that they diverged from the lineage leading to modern humans roughly 600,000 years ago. What seems more likely is that the theological humans leaving Africa succumbed to a level of bestiality, a behavior not altogether surprising given these humans would have been living in a fallen world. The fall and its relationship to the evolutionary origin of humans is discussed in more detail in chapter 9.

MITOCHONDRIAL EVE: THE GENETIC RECORD

Reconstructing the story of human origins and the emergence of modern human behavior is difficult given the different interpretations that can be given to the data. For example, there is robust debate over the nature and importance of the cultural artifacts associated with both modern humans and Neanderthals. Given this, it's likely that answering the question of when our prehuman hominin ancestors first gained that divine spark and became theologically human is beyond our ability to reconstruct.

What we do know is that all modern humans are genetically

related and that the most recent common ancestor of all living humans lived in the not-so-distant past. In fact, the dating of this ancestor gives us some idea of the minimum age at which the first theological humans emerged. Using mitochondrial DNA, which is only passed down through the female lineage, researchers have been able to approximate the age of the most recent common female ancestor of all living humans. This prominent figure, popularly referred to as Mitochondrial Eve, is thought to have lived roughly 150,000 to 200,000 years ago.[36] While many Christians view this research as a vindication of the Adam and Eve creation story, such a reading is a misinterpretation of the research.

If you take any two people and trace back their ancestry through their mother's line—take each person's mother, then their mother's mother, and then their mother's mother's mother, and so on—one will eventually find a woman who is the common female ancestor of both lineages. If all theological humans are related to each other by common descent (as both evolutionary theory and Catholic teaching would claim), it follows that there must be a woman who is the mother's mother of us all. Comparing the mitochondrial DNA sequences from all living ethnic groups and extrapolating from known rates of change in the mitochondrial DNA, researchers have been able to approximate when this most recent common female ancestor of all living modern humans existed—again, roughly 150,000 to 200,000 years ago.[37]

To be clear, this Mitochondrial Eve is not necessarily the first woman who existed. If Mitochondrial Eve lived 150,000 years ago in a small human population, she may have been one of a hundred women living at the time. It is possible that the female descendants of all the other women died off after a number of

36. E. Callaway, "Genetic Adam and Eve Did Not Live Too Far Apart in Time," *Nature* (2013), https://doi.org/10.1038/nature.2013.13478.

37. Krzysztof A. Cyran and Marek Kimmel, "Alternatives to the Wright–Fisher Model: The Robustness of Mitochondrial Eve Dating," *Theoretical Population Biology* 78, no. 3 (2010): 165–172, https://doi.org/10.1016/j.tpb.2010.06.001.

generations such that Mitochondrial Eve was the only woman who left female descendants in these subsequent generations. Such a scenario is not unusual, as studies of English surnames, which indicate the paternal lineage much like the Y chromosome would, have found that many of these surnames go extinct over time.[38] A similar situation occurs in family lineages, as one generation may not produce either a male or female offspring that has children. This would cause either the paternal or the maternal lineage to come to an end in this particular family lineage. Such a scenario is even more likely to occur in prehistoric times given the higher incidence of famine and disease outbreaks. It should be apparent then that what researchers call Mitochondrial Eve, the most recent common female ancestor of us all, is quite distinct from what Christians traditionally refer to as Eve, the first woman God created.

Similar research has been done to identify the most recent common male ancestor using Y chromosome data, as the Y chromosome is only passed down along the male lineage. This so-called Y-chromosomal Adam is also thought to have lived roughly 150,000 to 200,000 years ago.[39] The exact dating of both Mitochondrial Eve and Y-chromosomal Adam continues to be the subject of much debate, and it would not be surprising if new genetic information leads to some minor revisions of these dates. It is important to note that while the current estimated dates are similar, they do not necessarily have to be, because these studies are not looking for a first couple, the primordial Adam and Eve. Rather, they are looking for two distinct, unrelated individuals, the father's father of us all and the mother's mother of us all, both

38. Jasper Cropping, "Traditional Surnames Are Becoming Extinct: Farewell to the Footheads and Pauncefoots," *The Telegraph*, November 18, 2002, https://www.telegraph.co.uk/news/newstopics/howaboutthat/9685356/Traditional-surnames-are-becoming-extinct-farewell-to-the-Footheads-and-Pauncefoots.html.

39. G. David Poznik et al., "Sequencing Y Chromosomes Resolves Discrepancy in Time to Common Ancestor of Males Versus Females," *Science* 341 (2012): 562–565, https://doi.org/10.1126/science.1237619.

of which are individuals who presumably lived in an early human population in Africa, although not necessarily at the same time.

Taken together, the genetic, fossil, and archaeological evidence all indicate that modern humans have been around for some time. From a Catholic perspective, our common female ancestor, Mitochondrial Eve, and our common male ancestor, Y-chromosomal Adam, must have been theologically human. This is necessary to ensure that all their descendants (all living humans) are theologically human. If this wasn't the case, it would produce a scenario in which modern humans descended from different groups of non-theological humans; for example, Asians emerged from one population and Europeans emerged from another unrelated population. Such a polygenic origin is problematic from a Catholic perspective, as the Church teaches the unity of the human race, particularly as it relates to the concept of original sin, which is "transmitted by propagation to all mankind."[40] However, if the genetic evidence is accurate, then one could posit that theological humans emerged around 150,000 years ago in Africa and that all living humans descended from this population.

WHAT DOES THIS ALL MEAN?

This above scenario, that modern humans emerged in Africa relatively recently, is the common view held by most scientists. There are many other caveats and unknowns regarding how exactly modern humans emerged from other hominin populations and eventually spread throughout the globe, but their origins in Africa fit well with the available genetic and paleontological evidence. Such a scenario has also been addressed by the Catholic Church via the work of the International Theological Commission (ITC). The ITC is a group of Catholic theologians who advise the Church on theological matters, and this

40. *CCC* 404.

commission discussed human evolution in a 2004 document entitled "Communion and Stewardship: Human Persons Created in the Image of God," a work approved by then–Cardinal Joseph Ratzinger. The document declares that "while the story of human origins is complex and subject to revision, physical anthropology and molecular biology combine to make a convincing case for the origin of the human species in Africa about 150,000 years ago in a humanoid population of common genetic lineage. However it is to be explained, the decisive factor in human origins was a continually increasing brain size, culminating in that of *homo sapiens.*"[41]

To be clear, the authors are referring to the decisive *evolutionary* factor in the origin of the human form when they refer to "increasing brain size." From a Catholic perspective, the true decisive factor is the creation *ex nihilo* of an immaterial rational soul. Regardless of how the physical human form with its increased brain size and smaller jaw came to be, only an ensouled body is a human person. Without the immediate creation of the human soul, there is no such thing as a human person. Evolution alone cannot explain this act, as the ITC document notes: "Catholic theology affirms that the emergence of the first members of the human species (whether as individuals or in populations) represents an event that is not susceptible of a purely natural explanation and which can appropriately be attributed to divine intervention. Acting indirectly through [evolution], God prepared the way for what Pope John Paul II has called 'an ontological leap' . . . the moment of transition to the spiritual."[42]

The Genesis text describes how "God formed man from the dust of the ground, and breathed into his nostrils the breath of life" (Gen. 2:7). This line beautifully captures the physical and spiritual unity of man. One can see in the evolutionary process

41. International Theological Commission, "Communion and Stewardship: Human Persons Created in the Image of God" 63, 2002, vatican.va.

42. "Communion and Stewardship" 70.

"the dust of the ground" being appropriated for the emergence of man. This, though, is not enough. There is also the life of God that is breathed into man to animate his inner spiritual reality, his immaterial soul. This is the leap that separates theological humans from other creatures. This is the leap that John Paul II refers to as the divine creation of a human person, the only creature made in the image and likeness of God. As John Paul II implies, Catholics see a divide between humans and the rest of creation. Humans stand out from other created entities in terms of both our God-given abilities and our divine destiny.

While this distinction between humans and the rest of creation is obvious to anyone who observes human behavior, it can become blurred as one examines the archaeological and fossil record, as it is nearly impossible to pinpoint exactly when and where symbolic and conceptual thinking emerged in the evolutionary past. This makes it difficult to answer the question of when in the evolutionary process the first theological human person came into existence. For Catholics, the evolutionary emergence of man raises other sticky questions that are also difficult to answer definitely. Did the first theological humans emerge as a single couple, the proverbial Adam and Eve, or did they emerge as a population? How does one incorporate the mysterious reality of original sin into a scenario involving the evolutionary origin of the human body? These are questions we turn to in the next chapter.

9

Original Sin and the Evolution of Man

Certainly nothing offends us more rudely than this doctrine; and yet, without this mystery, the most incomprehensible of all, we are incomprehensible to ourselves.[1]

—Pascal

As Blaise Pascal highlights, there is something paradoxical at the heart of the Catholic doctrine of original sin. On the one hand, it seems patently unjust that we should be subject to the effects of a sin that we did not personally commit. It is difficult enough for us to accept responsibility for our own actions; to be held accountable for the actions of our distant forebearers seems a bridge too far. On the other hand, the claim that there is a deep wound that runs through the depths of our common humanity seems all too evident by the ubiquitous death and destruction that marks the whole of human history.

Even a cursory glance at the world around us provides ample evidence that something is amiss. An honest introspection of our own actions and motivations reveals a similar conclusion. There seems to be "a wound which is present in man's inmost self,"[2] and it is a wound that, for all our striving, we seem to lack the ability to overcome. In our more candid moments, the doctrine of original sin seems to be the only way to make sense of the disconnect that St. Paul so eloquently expressed in his Letter to the Romans:

1. Pascal, *Pensées*, section 7, fragment 434 (New York: E.P. Dutton, 1958), 121.

2. Pontifical Council for Justice and Peace, *Compendium of the Social Doctrine of the Church* 116, June 29, 2004, vatican.va.

"I do not understand my own actions. For I do not do what I want, but I do the very thing I hate" (Rom. 7:15).

Yet, for all its ability to speak to the reality of our shared human condition, the concept still does not sit well with our modern sensibilities. This is the mystery of original sin to which Pascal refers. It speaks truth to the human condition and to the struggles that occur within the human heart, but it seems hopelessly out of touch with modernity. It is certainly foreign to our legal understanding of justice to be held accountable for the actions of others. It is even more incomprehensible to those moderns who view man as a purely material being. The notion of a state of original justice followed by a subsequent loss of sanctifying grace and preternatural gifts seems utterly fanciful to those who view man as nothing more than a lump of matter.

In addition to these difficulties, an evolutionary understanding of the emergence of man seems to render the traditional understanding of original sin obsolete. The evidence for the emergence of man from other hominin populations seemingly contradicts the Genesis story of a primordial human couple, Adam and Eve, who were created in a state of original justice and holiness and who subsequently lost these gifts through sin. Absent this first "untainted" couple, a scenario that doesn't fit neatly into an evolutionary worldview, the Catholic dogma of original sin seems difficult to comprehend. In fact, some theologians have simply jettisoned the concept altogether, explaining away "original sin" as merely the disordered tendencies toward anger, greed, and lust that we have inherited from our primate ancestors. Viewed in this manner, there was no fall, nor was there an original state of justice, because we humans were fallen from the very start. Within the Catholic tradition, this is not a viable option.

How then does one reconcile original sin with an understanding of the science of human evolution? Is there any hope of integrating the two? To begin such an attempt, it is useful to put the Genesis account of the fall into the appropriate context and to tease out what the Church actually teaches about the fall and original sin. Because conceptions of original sin are often associated with a literal/historical understanding of the Adam and Eve story, original sin can be seen as inherently in conflict with the evolutionary history of man. In addition, the depiction of the Adam and Eve story in both childhood Bible-based storybooks and some classical artwork tends to facilitate the view that the Garden of Eden was a fantastical utopia that contained no pain or suffering for all of God's creatures. While this view tends to be common among Christians, such a view is not part of Church teaching. Likewise, it is important to recognize that the Genesis text of the first creation account merely states that creation is good, not perfect.

Yet there are those who suggest that before the fall there was no death or destruction in the world (a scenario that contradicts the evidence from evolutionary biology)—that the lion literally lay down with the lamb, or that man sustained himself by eating only plants in the Garden of Eden. While some of these interpretations might seem consistent with a literal/historical reading of the Genesis text, they seem inconsistent with the data indicating that man emerged from an evolutionary process that is "red in tooth and claw." However, much of this apparent conflict flows from biblical interpretations that do not represent Church teaching on the matter and are instead the result of cultural associations the Adam and Eve account has acquired over time. Therefore, it is critical to identify what the Catholic Church actually teaches regarding the fall.

WHAT DOES THE CHURCH TEACH ABOUT THE FALL?

The *Catechism of the Catholic Church* is quite clear that one cannot merely take the account of the fall in Genesis as an actual historical depiction of events. The *Catechism* states that "the account of the fall in Genesis 3 uses figurative language but affirms a primeval event, a deed that took place *at the beginning of the history of man*."[3] Given that it is a figurative account of a primeval event, one does not necessarily have to assume that there was a talking serpent or a tree of the knowledge of good and evil or that God banished our first parents from some earthly paradise, even though venerable Church Fathers such as Augustine tended to believe this was the case. However, even in the case of Augustine, who advocated for and defended a predominately historical reading of the fall, there was an openness to a nonhistorical reading if that became necessary based upon the evidence. In his *Literal Meaning of Genesis*, he stated the following: "Of course, if it became utterly impossible to safeguard the truth of the faith while accepting in a material sense what is named as material in Genesis, what alternative would be left for us except to take these statements in a figurative sense rather than to be guilty of an impious attack on Sacred Scripture?"[4]

However, if the language is to be best understood figuratively, the *Catechism* is clear that the sacred author used this depiction to affirm "a primeval event," something that actually occurred at the dawn of humanity. It depicts the first rejection of God by mankind and points to the disruption in relationships that this event has caused. The Church does not make any particular claim regarding the nature of the first sin (although the wider tradition characterizes it as the sin of pride) but simply holds that it occurred at the beginning of human history.

The idea that the first humans, those who first had the capacity to know and freely love God, would have succumbed to temptation hardly shocks our sensibilities given our intimate familiarity with human failings. In fact, that there would be a first sin is nothing more than stating the obvious (if one agrees with the reality of sin). If human sin is the result of knowingly

3. *Catechism of the Catholic Church* 390 (emphasis added).

4. Augustine, *The Literal Meaning of Genesis* 8.1.4, trans. John Taylor (New York: Newman, 1982).

rejecting God, once human creatures emerged with the spiritual capacity to freely embrace or reject the Creator, that first rejection was immediately a possibility. Indeed, it is a possibility that our first parents actualized. How quickly our first parents turned from God is unknown, although St. Maximus the Confessor suggested that "our nature unnaturally fell at the instant it was created."[5] The fact that this first sin occurred (however rapidly) is not what is so unsettling about the doctrine of original sin. Rather, it is the next line of the *Catechism* that is harder for many to digest: "Revelation gives us the certainty of faith that the whole of human history is marked by the original fault freely committed by our first parents."[6]

How exactly does this first transgression mark all of human history? To unpack this, it is important to recognize that sin is always something that goes beyond the individual. According to Joseph Ratzinger, "Sin is loss of relationship, disturbance of relationship, and therefore it is not restricted to the individual. When I destroy a relationship, then this event—sin—touches the other person involved in the relationship. Consequently, sin is always an offense that touches others, that alters the world and damages it."[7]

The first humans were invited into a special relationship with God and were given the grace of original holiness and justice. When they lost this gift through sin, they rejected the intended relationship they were to have with God. In so doing, they also rejected the relationships God intended them to have with their fellow men and with creation in general. The key is that these right relationships are only possible through the grace of God. Humans were created to be dependent upon God's grace for their

5. Maximus the Confessor, *Ambiguum* 42, PG 91, 1321B, in *On the Cosmic Origins of Christ: Selected Writings of Maximus the Confessor*, trans. Paul Blowers and Robert Louis Wilken (Crestwood, NY: St. Vladimir's Seminary Press, 2003), 85.

6. *CCC* 390.

7. Joseph Ratzinger, *"In the Beginning . . .": A Catholic Understanding of the Story of Creation and the Fall* (Grand Rapids, MI: Eerdmans, 1995), 73.

very fulfillment. With the rejection of God through sin, that grace of original holiness was lost, and mankind was left with the physical body evolution had produced, one no longer in a fully harmonious union with the rational soul. As the *Catechism* states, "The harmony in which they had found themselves, thanks to original justice, is now destroyed: the control of the soul's spiritual facilities over the body is shattered. . . . Harmony with creation is broken."[8]

This damaging of relationships is then transmitted to subsequent generations in a manner that "is a mystery that we cannot fully understand."[9] What is understood, though, is that original sin is a loss of something (sanctifying grace and preternatural gifts) rather than a personal sin by the individual: "It is a sin 'contracted' and not 'committed'—a state and not an act."[10] This leads to the apparent paradox that original sin is simultaneously foreign to us and part of us. While the act was our parents', it is ours in that we live deprived of the preternatural gifts and inhabit a world that is infected by the reality of sin and marked by a lack of that original holiness. That is *our* condition.

An analogy might be useful to help explain this concept. If my parents were given a gift of a large fortune but in turn squandered it on questionable investments, that fortune would be lost such that they could no longer hand it down to their offspring. My life would be affected by the lack of this gift. In an analogous fashion, that original state of holiness was squandered such that it could not be passed on. That gift, rejected by our first parents, is no longer present. Rather, what is passed on is "a human nature deprived of original justice and holiness."[11] As a result, the effects of that sin continue to adversely affect our relationships with God, with each other, with ourselves, and with creation. On our own,

8. *CCC* 400.
9. *CCC* 404.
10. *CCC* 404.
11. *CCC* 404.

we cannot rectify the situation. As Ratzinger stated, "Only God's love can purify damaged human love and radically reestablish the network of relationships that have suffered alienation from sin."[12]

This doctrine of original sin does a good deal of heavy lifting within Catholic theology. It explains how sin has infiltrated a world that God, who is infinite goodness, created good. It is humans who, through our freedom to reject God, brought sin into the world. Yet original sin also points to the Good News and our need for Christ's redemption. If reliance upon our own abilities has ushered in sin and death, reliance upon the grace of God through Christ's sacrifice can usher in our transformation to a new and higher state of justice and holiness. Original sin is then intimately tied to these key principles of Catholic theology:

1. God, in his infinite goodness, created a world that is good. However, the "universe was created 'in a state of journeying' (*in statu viae*) toward an ultimate perfection yet to be attained."[13]
2. Humans have a supernatural destiny; we are meant to experience the beatific vision and love of God in heaven.
3. Sin entered our world through human action and subjected the entire human race to its effects.
4. Humans, as created beings, lack the ability on our own to restore fully the relationships we have damaged.
5. Christ's sacrifice is necessary to repair fully these relationships and to ultimately restore our divine sonship and heavenly inheritance.

With these principles in mind, it is worth examining in more detail the way some of the Fathers of the Church have described original sin and the fall. This will help clarify these positions

12. Ratzinger, *"In the Beginning,"* 74.
13. *CCC* 302.

within the tradition and set the stage for an examination of how these key principles might be integrated with evolutionary theory.

ORIGINAL SIN AND THE CHURCH FATHERS: AUGUSTINE AND IRENAEUS

Among the Church Fathers, original sin finds its first full description in the writings of St. Augustine in the fourth century. While the concept had been discussed prior to Augustine, writers who had done so, such as Justin Martyr, discussed the concept without making any reference to the term "original sin." Augustine, for his part, devoted a considerable amount of effort in his writings to the proper interpretation of the Genesis text. While his interpretation of the account of the fall was primarily historical, there were certainly instances in which he viewed the text of Genesis as symbolic or allegorical. He explained his overall approach to interpreting the Genesis text in his work *On the Literal Interpretation of Genesis*: "One may expect me to defend the literal meaning of the narrative as it is set forth by the author. But if in the words of God, or in the words of someone called to play the role of prophet, something is said which cannot be understood literally without absurdity, there is no doubt that it must be taken as spoken figuratively in order to point to something else."[14]

As a result, he did not equate the "days" of creation with literal twenty-four-hour days given that the heavenly bodies that define the length of a day do not appear until day four in the text. However, when it came to the account of the fall, he believed that the text suggested a literal/historical interpretation. As a result, he believed that one should take the description of Satan joining his spirit to the serpent, the eating of the forbidden fruit by Adam and Eve, and the banishment from the garden as actual historical events. Given the worldview of the time and the absence of any scientific evidence or historical scholarship to the contrary,

14. Augustine, *Literal Meaning of Genesis* 11.2.

it is not surprising that he interpreted the Genesis account in this manner. However, he was open to the possibility that certain aspects of the story could be seen as figurative descriptions, even though he did not take them as such. In particular, he acknowledged that to "understand the tree not in the proper sense as a real tree with real fruit but in a figurative sense . . . could result in a theory apparently consistent with faith and reason."[15]

Augustine viewed the fall or original sin as a cataclysmic event that altered all of creation. He saw mankind as marked by Adam's sin and damned by the inheritance of it. This sin was passed on via generation as a corrupt, fallen version of human nature that one inherited from one's parents. He believed that our human nature had been radically transformed by the sin of Adam such that our free will was greatly diminished and that humanity had been reduced to a "mass of perdition" through its effects. It is worth noting that Augustine's thoughts on original sin were formed during the Pelagian controversy and stand in stark contrast to Pelagius' views. Pelagius argued, among other things, that Adam's sin only affected Adam, that original sin represented a bad example and no more, that original sin is propagated by imitation, and that infants do not need Baptism. Augustine opposed all of these positions, and his views on original sin make this clear.

While Augustine's view of original sin has had the most influence on Christian theology throughout the centuries, other Fathers developed slightly different views of the concept. Irenaeus' approach to original sin is one such example. Irenaeus emphasized creation as being in a state of development and growth toward a greater perfection. In this manner, he viewed Adam and Eve as having been created in a state of youthfulness or immaturity rather than being created as mature adults. According to Irenaeus,

15. *Literal Meaning of Genesis* 11.56.

"Man was a young child, not yet having a perfect deliberation, and because of this he was easily deceived by the seducer."[16]

Despite these differences in how Augustine and Irenaeus viewed the fall of Adam of Eve, both viewed it as a historical event. However, Irenaeus was less interested in the details of the event and the transmission of original sin than Augustine. In fact, he never developed the concept of original sin in a similar fashion, focusing rather on the role Christ had in restoring mankind to the original state of sinlessness. In his work *Against Heresies*, he stated, "Thus, because it was not possible for that man who had once been conquered, and thrust out by disobedience, to be new molded and obtain the prize of victory, and again it was impossible for him to obtain salvation, who had fallen under sin: The Son accomplished both, being the Word of God, coming down from the Father, and made flesh and descending even unto death and fulfilling the Economy of our salvation."[17] On this point, the necessity of Christ for our salvation, Irenaeus agreed with Augustine. In fact, despite the differences found in their accounts on original sin, both Augustine and Irenaeus, like the rest of the Church Fathers, agreed on the five principles related to the doctrine of original sin described in the previous section.

THE ISSUE WITH EVOLUTION

For the Fathers, the assumption that Adam and Eve were created directly by God *ex nihilo* provided both a clear manner by which man could have been formed in a state of original justice and an explanation of the unity of the whole human race. This historical interpretation of the text, which the Fathers had little reason to question given the scientific understanding of their times, could easily incorporate the belief that the effects of the fall (the loss of

16. Irenaeus, *On the Apostolic Preaching* 12, trans. John Behr (Crestwood, NY: St. Vladimir's Seminary Press, 1997), 47.

17. Irenaeus, *Against Heresies* 3.18.2, trans. John Keble (Oxford: James Parker, 1872), 275.

the original state of holiness and justice) were universal, in that this loss was propagated to all of Adam and Eve's descendants (a position that was taught by the Council of Trent).[18]

The advent of evolutionary theory, particularly the evidence suggesting that the human physical form evolved from other hominins over time, brought this neat and tidy interpretation into question. If man had evolved from other hominins, how did man emerge into the world in a state of original justice? Did not man's sin bring death into the world? Why then does the scientific evidence display signs of death and destruction long before man's arrival?

While the Church teaches that man's sin severed our relationship with the Creator such that we became subject to both natural and spiritual death, it does not teach that creation was in some blissful state where the lion lay down with the lamb before the fall. Such a belief is often attributed to St. Augustine, who saw the fall as a catastrophic event that radically ruptured the relationship between man, the Creator, and creation. However, it seems clear from his writings that he did not hold to a view that there existed a prelapsarian earthly paradise in which other animals experienced no death or predation before the fall of man. Even in the case of man, Augustine made a distinction between man being created with the possibility of not dying (a position he held) and man being created in a state in which death was impossible. Regarding man's "immortality," he states, "It is one thing, after all, not to be able to die, like natures which God created immortal, while it is quite another to be able not to die; and this is the way the first man was created immortal, something to be granted him . . . not by his natural constitution."[19]

In fact, for Augustine, the beings God created were potentially subject to corruption from the beginning, not because of

18. General Council of Trent, Fifth Session, Decree on Original Sin, June 17, 1546, in *The Sources of Catholic Dogma*, ed. Henry Denzinger and Karl Rahner, trans. R.J. Deferrari (St. Louis, MO: Herder, 1954), 247.

19. Augustine, *Literal Meaning of Genesis* 6.35.

any fault in the beings themselves, but rather merely from the fact that they were created beings. As created beings, he considered them fundamentally contingent and open to corruption and change. As a result, Augustine did not see animal creatures as being in some prelapsarian state of perfection before the fall and then transforming post-fall into dangerous and deadly creatures. He addressed this specifically in book 3 of *The Literal Meaning of Genesis*, where he used the examples of Daniel in the lion's den and the case of the deadly viper clinging to St. Paul's arm. In these cases, the viper and the lion still possess their deadly powers, but they inflict no harm upon St. Paul and Daniel. An analogous situation would have occurred before the fall. According to the theologian Stan Rosenberg, "The viper, a poisonous snake, is described [by Augustine] as good; it is because of something wrong in humans that we find such creatures to be a problem. Vipers, in other words, did not change; they possessed poison and the ability to kill as part of their original form. Rather, what changed was humans' relationship to them. Augustine nowhere suggests that the animal world underwent a biological transformation after the fall."[20]

While the Church teaches that man became subject to physical death after the fall, it is the possibility of man's spiritual death that is of paramount importance. With that first sin came the possibility of spiritual death, an eternal consequence for choosing to reject a life with God. This possibility for spiritual death could not exist until 1) spiritual creatures (humans) were brought into existence and 2) these creatures then separated themselves from God via a deliberate choice.

Yet, for human death to be the consequence of a deliberate human choice, humans must have come into being in a state of original justice in which they were not subject to death, spiritual

20. Stanley P. Rosenberg, "Can Nature Be 'Red in Tooth and Claw' in the Thought of Augustine?" in *Finding Ourselves after Darwin: Conversations on the Image of God, Original Sin, and the Problem of Evil*, ed. Stanley P. Rosenberg (Grand Rapids, MI: Baker Academic, 2012), 237.

or otherwise. But that raises the question: How did humans emerge evolutionarily in a state of original justice from other hominins who had bodily desires that from our perspective seem disordered? To address this sticky question, some theologians have suggested that original sin should be viewed merely as the fact that, through the evolutionary process, humans have inherited from other hominins the natural tendencies toward aggression, lust, and selfishness. While this may make sense from an evolutionary perspective, the inheritance of such disordered desires would seem to make humans flawed from the start, a position clearly at odds with the doctrine of original sin.

How then did we gain these disordered desires that seem to be shared with other primates? Acknowledging that we have inherited biological drives from our hominin ancestors, drives that we know from experience can become disordered, is not something that necessarily undermines the Catholic understanding of original sin. While these biological drives were inherited, one must recognize that the first humans were blessed with preternatural gifts that allowed them to live in a state in which these drives were perfectly ordered toward the good. This is the state of original justice. This was not a state in which the biological desires we share with other primates did not exist; rather, it was a state in which they were properly controlled. Man could not order these on his own but was dependent upon his Creator for this gift. The fall then represents the loss of this gift, leaving humans in a postlapsarian state with a nature that was then at the mercy of these biological drives *without the aid of the original grace God had intended.* Left to our own devices—namely, the animal nature inherited from other primates and our rational souls—these drives and biological urges, which are amoral in nonrational animals, fail to remain properly under the control of the intellect. This is the state in which we find ourselves, saddled with the burden of attempting to order our desires without the aid of that original gift of divine grace. Original sin is the name of this condition.

When speaking of human nature, it is worth reiterating what is meant by this term. From a Catholic perspective, human beings are seen as a unity of a physical body (matter) and an immaterial rational soul (spirit). Echoing Aquinas, the *Catechism* states that "spirit and matter, in man, are not two natures united, but rather their union forms a single nature."[21] This one human nature, the unity of an evolved body and a spiritual rational soul, is what we all possess. What we currently lack is the original grace that allowed us to rightly order that unity. Evolutionarily, the first humans would have emerged from a distinct nonhuman hominin population, a population of individuals of a different kind. (They possessed a different body and they lacked a rational soul.) Exactly how this might have occurred is not clear (and likely never will be), but speculating on the manner can bring some clarity to the theological issues involved.

HUMANS FROM HOMININS

The origin of the first human is an event of direct creation given that the human soul is created immediately by God. At the same time, though, the first human is the result of the evolutionary process, which, operating within God's providence, is thought to have produced the type of entity that was fitting to receive a rational soul. In this sense, man is both the product of evolution and the product of the direct and immediate creation by God. This is analogous to how a new human life is the result of both the material contributions of the mother and father (the sperm and ovum) and God's immediate creation of the human soul.

While humans provide the material contribution for new humans to come into being, who or what provided the material contribution for the first human? From an evolutionary

21. *CCC* 365.

perspective, this would have been provided by an advanced hominin couple. In this scenario, the hominin parents would have provided specific novel material changes (genetic alterations in a sperm and/or an oocyte or changes to the composition of the oocyte) such that the specific fertilized egg was the type of entity that was fitting to be endowed with a rational soul. Of course, this first human spiritual soul, precisely because it is spiritual, would have been created immediately by God, as it is *not* the type of entity that can evolve.

In such an event, the emergence of the first human person would have occurred within the womb of a nonhuman hominin. While the ensoulment of this child would have been a supernatural event, it would not be considered miraculous. This first human zygote, like all subsequent human zygotes, would be the type of entity that is always, by God's direct action, an ensouled human being from the very moment of conception. (A chimpanzee or nonhuman hominin zygote is not.) At some point in the hominin lineage, enough physical changes may have occurred in the gametes of the mother and father such that the zygote that was conceived was the type of entity that was fitting to be endowed with an immaterial rational soul. It would become the first human person. In addition, this first human would also have been gifted with supernatural graces such that his entire person, body and soul, would be ordered rightly toward human flourishing.

It may seem odd or unsettling that these first humans would have been born of and raised by nonhuman, highly advanced hominin parents. Such a scenario was disconcerting to the Catholic anti-evolutionist Ernesto Cardinal Ruffini, who wrote, "Who will be ready to believe that Adam had for his father and mother two brute beasts? Although innocent and holy, he would certainly have been in a much less honorable condition than is ours."[22]

22. Ernesto Ruffini, *La Teoria dell'evoluzione secondo la scienza e la fede* (Norcia: Orbis Catholicus, 1948), trans. Francis O'Hanlon, in *The Theory of Evolution Judged by Reason and Faith* (New York: Wagner, 1959), 139.

Despite Ruffini's concerns, though, the Church has not discussed this possibility in any official capacity, and such a scenario does not seem to contradict any dogmatic teachings of the Church regarding the nature and origin of man.[23]

While this scenario may seem strange, given the likely biological similarities between the two, it seems hardly a stretch to believe that advanced hominin parents could have physically provided for and nurtured the first humans. What such parents could not have been able to provide is the cultural learning, such as language acquisition, that is part and parcel of our shared human existence. Despite this lack, given that the Church teaches that these first humans would have been endowed with preternatural gifts (infused knowledge, etc.), they may have been able to develop properly these innate human abilities in ways we cannot conceive.

While some may recoil from the idea that the first humans were conceived in the wombs of advanced hominins, it is worth contemplating that two analogous events stand at the center of our faith: the Son of God being conceived in the womb of a mere creature and Mary being conceived without sin in the womb of a sin-stained mother. In fact, when properly viewed in light of the Incarnation (the new Adam) and the Immaculate Conception (the new Eve), the evolutionary origin of Adam and Eve in the womb of advanced hominins seems quite fitting. This is particularly the case given how the typology of the Old Testament is always surpassed by that which is found in the New Testament.

In fact, it would be hard to argue that humans emerging from the wombs of advanced hominins is more scandalous than the belief that God emerged from the womb of Mary. Clearly, God could have chosen any manner of ways of entering our world. Christ could have entered the world as a fully formed thirty-year-old Galilean who set out to preach the Gospel. Yet

23. Kenneth Kemp, "God, Evolution, and the Body of Adam," *Scientia et Fides* 8, no. 2 (2020): 139–172.

he chose to come to us as all men do—as a zygote in the womb of a creature—as a sign that he came to redeem all aspects of our humanity.

It seems that the typology of Christ as the new Adam could take on an even deeper level of meaning if such an evolutionary scenario were true. In fact, such a scenario creates interesting parallels between Christ's Incarnation and man becoming incarnate in the womb of a hominin. In the case of the former, Christ did not just cursorily enter creation fully formed to set about his work of redemption. Rather, he entered it completely, taking part in every stage of human development. In doing so, he allowed himself to be dependent upon humans—namely, Mary and Joseph—for his physical needs and survival. He became like us in all things but sin so that he could redeem us completely.

Likewise, the first humans would not have entered creation fully formed in a disruptive manner, much like a foreign interloper. Rather, they would have entered and participated in every stage of human development. In doing so, they would have been dependent upon other creatures for their physical survival just as Christ made himself dependent upon Mary and Joseph. But just as Christ's divinity was not compromised or diluted by being nurtured by mere humans, the first humans would not have had their human nature compromised by being nurtured by advanced nonhuman hominins. (This is particularly true given that they would have been endowed with the original state of justice and holiness.)

In fact, given the Catholic notion that man is the crown of creation, it would seem fitting that the first man and woman be brought to completion with the direct aid of God's creation, a creation in which we all find ourselves deeply embedded and for which we are given responsibility. Just as Christ, who is the fulfillment of humanity, took his material nature from Mary and transformed it, the first humans, who are the fulfillment of creation—the only creature God willed for its own sake—would take their material nature from hominins such that God, through the creation of the human soul, could transform it.

While deep theological parallels with the new Adam and the new Eve exist with this scenario, it is clearly speculative, as the Church has not provided guidance on this issue. As such, there are certainly other evolutionary scenarios one could posit by which the first humans could have emerged. Possibly the first humans did not become rationally ensouled until they were older. Much like Aquinas believed humans went through stages *in utero* in which they were informed by various types of souls, maybe the first humans were informed by an animal soul until the age of reason, at which point they became informed by a rational human soul. This scenario raises other issues, though, such as the difficulties of explaining why the same entity would be informed by different souls at different stages of his or her life.[24] In addition, why would this delayed ensoulment only occur at the dawn of humanity and not continue down through subsequent generations?

Regardless of the scenario one posits, though, one must recognize that the first humans were given supernatural aid from the outset. Due to being endowed with preternatural gifts, the biological urges they inherited from their prehuman ancestors would have been ordered perfectly toward the good. Evil enters the story of human existence only after our first parents knowingly rejected this gift and thus allowed themselves to become subjugated to the biological urges present in ourselves and other, amoral animals.

MONOGENISM VS. POLYGENISM: WHAT THE SCIENCE SUGGESTS

When discussing original sin and human origins from a Catholic perspective, the notion that all humans have descended from a

24. Aquinas' notion of delayed ensoulment during human development *in utero* is discussed briefly in chapter 7. In this case, delayed ensoulment would have to expand beyond the womb, and rational ensoulment would occur in adolescence or early adulthood.

single couple, the primordial Adam and Eve, is usually the default assumption. This notion, known as monogenism, has strong roots in Catholic tradition and in Scripture. In his letters, St. Paul makes extensive use of the Adam-Christ typology to emphasize how Christ's sacrifice overcame the damning effects of Adam's sin. In addition, the Church Fathers often refer to Adam and Eve as the original human couple, and the *Catechism* makes clear reference to a couple as our first parents as it discusses, among other things, "the [initial] harmony between the first couple and all creation."[25]

In contrast to monogenism, the science suggests that the first humans emerged via polygenism—that is, as a population rather than as a single couple.[26] Given that evolutionary population thinking is a relatively modern scientific concept, it is no wonder that there is little in the Catholic tradition that speaks to this possibility. Today, though, population thinking permeates evolutionary biology at all levels, not just at the level of the emergence of humans. Within evolutionary theory, it is *populations* of organisms that, over expanses of time and under certain conditions, can evolve into new species. While new traits emerge in individuals, they must spread through existing populations in order to become established. As different traits emerge, spread, and accumulate in a subpopulation, over time this subpopulation can transform into a novel species that is distinct from the initial population. During this transition, the lines between the initial population and the new emerging species (the subpopulation) can be blurred. While the two populations may develop distinct features, they may still

25. *CCC* 376.

26. Polygenism can have two different meanings. The meaning used here is an example of monophyletic polygenism and is the position that humans emerged once (monophyletic) during evolutionary history (in Africa) as a population (polygenism). There is also the position of polyphyletic polygenism, which does not have many adherents within the scientific community. This is the position that humans emerged multiple times (polyphyletic) during evolutionary history in different populations. For example, modern humans would have emerged in Africa, but another population of modern humans would have emerged separately in Asia, and still another in the Pacific Islands. The genetic evidence strongly suggests a monophyletic origin of humans; therefore, only monophyletic polygenism is discussed here.

exhibit limited interbreeding with each other. At the initial stages of divergence, clear and fast lines separating these two populations can be hard to identify in practice.

When it comes to the emergence of humans, the scientific evidence suggests that they emerged in a similar fashion: as a distinct *population* that evolved from other related hominins. There are multiple lines of scientific evidence that point in this direction. The most compelling one has to do with the amount of genetic diversity that exists in the current human population. It turns out that there is more genetic variability in the modern human population than one would expect if all modern humans descended from an original couple in the relatively recent past. If one assumes that the current human population can be traced back to an original couple that lived 150,000 to 250,000 years ago, then all the genetic diversity that exists today (beyond the limited amount found in that original couple) must have originated relatively recently and relatively quickly. Based upon the rates of genetic change that have been measured within human lineages, such a scenario seems highly improbable.

In general, populations that have descended from small founding populations tend to have low genetic variation. For example, most of the more than forty thousand Hutterites[27] in North America originated from a small founding population (eighty-nine founding members), and they have remained relatively genetically isolated since their original founding. As a result, they have a low amount of genetic diversity, such that the Hutterite population is prone to a variety of recessive genetic disorders. If humans had a founding population of just two, this would reduce the initial genetic diversity even more than that seen in the Hutterite population, a situation that would greatly increase the detrimental effects of inbreeding.

27. Hutterites are a communal ethnoreligious branch of Anabaptists that are predominately found in the northwest regions of North America. They form intentional communities in which the members are relatively genetically isolated from the general population.

In addition, the Hutterite population in North America exhibits an obvious genetic signature due to a lack of genetic diversity. (There is much less genetic diversity in this population than in the general population.) Likewise, if the human population emerged from just two individuals in the relatively recent past, it would have left a distinct genetic signature that is quite different than what one would expect if humans emerged from a larger population.

Using a variety of methods, one can estimate the minimum human population size that had to exist at some time point in the recent past to account for the current level of genetic variation found in the human population. While there are different methods for measuring this, one common method is to look at what are called single nucleotide polymorphisms (SNPs), which are single bases[28] in the human genome sequence that can vary from person to person. Since the rate at which mutations occur and get shuffled or recombined in the genome can be observed and measured, researchers can extrapolate backward to estimate how big the population would have had to have been in the recent past in order to account for the current distribution of SNPs currently found in the human population.

What has been shown repeatedly, using a variety of different methods and estimates, is that there seems to be too much genetic diversity in the current human population to be explained by a starting population of two.[29] In such a scenario, one would have to posit an inordinately high mutation rate or recombination rate to extrapolate the current genetic diversity back to a starting pool of two individuals. Most estimates suggest that the human population size during the past 200,000 years likely was never

28. There are four possible bases found at each site in the human genome: A, T, G, or C. At most sites in the human genome, the entire human population has the exact same base. However, roughly 1 in every 1,000 sites are polymorphic, meaning that some people have one base, for example A, while others have a different base, such as C.

29. A. Tenesa et al., "Recent Human Effective Population Size Estimated from Linkage Disequilibrium," *Genome Research* 17, no. 4 (2007): 520–526, https://doi.org/10.1101/gr.6023607.

lower than one to two thousand individuals.[30] As the biologist S. Joshua Swamidass[31] points out, "There seems to be strong evidence that our [genetic] ancestors never dip down to a single couple at a point when *Homo sapiens* first appear."[32] The point is that if the entire current human population emerged from just two individuals 200,000 years ago, such a scenario would have left a strong genetic signature that is quite distinct from what we currently see. If these two individuals had lived just 10,000 to 20,000 years ago, it would have left an even stronger genetic signature. Such a genetic signature is lacking in the human population.

MONOGENISM VS. POLYGENISM: WHAT THE CHURCH TEACHES

Given the scientific evidence suggesting that *Homo sapiens* emerged from a population rather than a single couple, it is worth investigating what the Church has stated regarding the compatibility of polygenism and Catholic teaching. The most explicit reference to the issue can be found in Pope Pius XII's 1950 encyclical *Humani Generis*. After articulating that discussions regarding the evolution of man's physical body from "pre-existing and living matter" are permissible, Pope Pius goes on to state the following about polygenism:

> When, however, there is question of another conjectural opinion, namely polygenism, the children of the Church by no means enjoy such liberty. For the faithful cannot embrace that opinion which maintains that either after Adam there existed

30. B.M. Henn et al., "The Great Human Expansion," *PNAS* 109, no. 44 (2012): 17758–17764, www.pnas.org/cgi/doi/10.1073/pnas.1212380109.

31. Swamidass makes the argument for the relatively recent existence of Adam and Eve, whom he argues are the genealogical ancestors but not the sole genetic ancestors of all living humans. This position is discussed briefly later in this chapter.

32. S. Joshua Swamidass, *The Genealogical Adam and Eve: The Surprising Science of Universal Ancestry* (Downers Grove, IL: InterVarsity, 2019), 102.

> on this earth true men who did not take their origin through natural generation from him as from the first parent of all, or that Adam represents a certain number of first parents. Now it is in no way apparent how such an opinion can be reconciled with that which the sources of revealed truth and the documents of the Teaching Authority of the Church propose with regard to original sin, which proceeds from a sin actually committed by an individual Adam and which, through generation, is passed on to all and is in everyone as his own.[33]

While Pope Pius XII states that polygenism is a conjectural opinion that faithful Catholics are not at liberty to embrace, it is important to examine his rationale for this proscription. The issue Pope Pius XII identifies is not with polygenism per se, but rather with the implications that polygenism would have for the doctrine of original sin. He claims that it is "in no way apparent" how polygenism can be reconciled with the Catholic understanding of original sin as articulated by the Council of Trent—that is, as originating from the sin of our first parents (a single couple) and then transmitted "by propagation" to all humanity.[34]

It is of particular interest that Pope Pius XII used the phrase "it is in no way apparent" rather than a stronger statement such as "it is impossible to reconcile" or "it is forbidden to speculate upon how they could be reconciled." In fact, the philosopher Dr. Kenneth Kemp has indicated that early drafts of *Humani Generis* contained stronger language regarding polygenism, and this softer language was deliberately chosen for the final version. Given the language used, many scholars have suggested that if one were able to reconcile polygenism with the Catholic doctrine of original sin, there would be nothing inherently problematic with polygenism.

33. Pius XII, *Humani Generis* 37, encyclical letter, August 12, 1950, vatican.va.

34. General Council of Trent, Fifth Session, Decree on Original Sin, June 17, 1546, in *The Sources of Catholic Dogma*, ed. Henry Denzinger and Karl Rahner, trans. R.J. Deferrari (St. Louis, MO: Herder, 1954), 247.

Pope Pius' successor Pope Paul VI used very similar language regarding polygenism in an address on original sin in 1966, stating, "It is evident that you will not consider as reconcilable with the authentic Catholic doctrine those explanations of original sin, given by some modern authors, which start from the presupposition of polygenism *which is not proved*."[35] While Pope Paul VI does not explicitly condemn polygenism, he does caution against specific explanations of original sin that have been informed by a belief in polygenism (explanations he does not fully describe). Just like Pope Pius' wording in *Humani Generis*, though, Pope Paul's statement seems to leave open the possibility that new "explanations" could be offered that might be able to reconcile polygenism with Catholic teaching on original sin.

In fact, recent Church documents seem to be more open to this possibility. It is worth noting that in the 2004 document "Communion and Stewardship," which was quoted in the previous chapter, the Vatican's International Theological Commission (ITC)—a commission composed of theologians from diverse schools and nations noted for their knowledge and faithfulness to the Magisterium—made reference to a first human *population* rather than exclusively referring to an initial couple. Again, when discussing human evolution, it states, "While the story of human origins is complex and subject to revision, physical anthropology and molecular biology combine to make a convincing case for the origin of the human species in Africa about 150,000 years ago in a *humanoid population* of common genetic lineage."[36]

However, it makes clear that "Catholic theology affirms that the emergence of the first members of the human species (whether as individuals or in populations) represents an event that is not

35. Paul VI, Address to Theologians at the Symposium on Original Sin, July 11, 1966, vatican.va (author's translation).

36. International Theological Commission. "Communion and Stewardship: Human Persons Created in the Image of God" 63, 2002, vatican.va (emphasis added).

susceptible of a purely natural explanation and which can appropriately be attributed to divine intervention."[37]

While the status of this document is not on the order of a papal encyclical, the fact that an ITC document approved by then–Cardinal Joseph Ratzinger speculates on a first human population is certainly worth taking seriously. This does not indicate that a first human population is easily reconcilable with Church teaching on original sin, but it does suggest that, to paraphrase Pope Pius XII from *Humani Generis*, research and discussions on the part of those experienced in both science and theology should take place regarding this subject.

ISSUES WITH RECONCILIATION

Integrating the science of human origins with the Catholic doctrine of original sin is not a simple task. Despite the difficulties, a variety of scenarios have been offered as possible solutions to the problem. One of the most interesting hypotheses has been put forward by the philosopher Kenneth Kemp. In crafting a possible solution, Kemp makes the distinction outlined in the previous chapter between "theological humans," humans endowed with a rational soul, and "biological humans,"[38] organisms that could in theory successfully interbreed with theological humans despite some physical differences and the lack of a rational soul (which is discussed to some extent in chapter 8).[39] The key is that the theological humans, despite their differences in nature from purely biological humans, still retained the ability to interbreed successfully with biological humans (prehuman hominins). This would be analogous to the manner in which

37. "Communion and Stewardship" 70.

38. The term *biological humans* includes prehuman hominins that were able to successfully reproduce with true theological humans.

39. K.W. Kemp, "Science, Theology, and Monogenesis," *American Catholic Philosophical Quarterly* 85, no. 2 (2011).

modern humans retained the ability to interbreed with Neanderthals after they had diverged into separate populations.[40]

In this scenario, the initial theologically human couple, endowed with the grace of original holiness, eventually turns from God and rejects his gifts of grace. The consequence of this rejection (original sin) is that all of their offspring would inherit a human nature devoid of the accompanying gift of original grace and holiness. Yet, the fallen offspring of this initial couple, due to having enough physiological similarities, would have been able to successfully interbreed with the large number of merely biological humans within their midst.

What then to do with the offspring of such mating events? Kemp proposes the possibility that the offspring of these interbreeding events might be the type of entity that it is fitting for God to endow with a rational soul. While there are questions regarding the specifics of how this might work from a genetic perspective,[41] given their rational abilities, the theologically human population would be expected to have a significant evolutionary advantage over their purely biologically human neighbors. Over time, the theological humans would be expected to displace the purely biologically human population. They would do this while simultaneously incorporating some of the genetic diversity that existed in the biologically human population from which they arose.

While such interbreeding events might seem distasteful, in a world subjected to human sin and the consequences of the fall,

40. As discussed in the previous chapter, the idea that modern humans bred with nonhumans is well supported by the available evidence. For example, sequencing data indicates that all non-African humans share about one to three percent of their DNA with Neanderthals, indicating that modern humans interbred with Neanderthals as they left Africa and encountered Neanderthals in the Levant, Asia, and Europe. See F.A. Villanea and J.G. Schraiber, "Multiple Episodes of Interbreeding between Neanderthal and Modern Humans," *Nature Ecology & Evolution* 3, no. 1 (2019): 39–44, https://doi.org/10.1038/s41559-018-0735-8.

41. Given the inherent stochastic nature of gamete formation, it may take several generations for the genetic factors associated with theological humans to become stabilized within the population.

such actions would be seen, in some sense, as the natural consequence of the introduction of that sin. As the *Catechism* states, "After that first sin, the world is virtually inundated by sin."[42]

While this type of bestiality may not be appealing as a hypothesis, incest seems to be the logical consequence of the traditional "Adam and Eve" scenario. If all genetically descended from Adam in the traditional manner, the only way to keep the population expanding on down to the third generation is to posit incestual relationships between Adam and Eve's offspring.[43] In both cases, bestiality or incest, the sexual deviancies would be consistent with a fallen state of human existence. In the former case, though, one has the benefit of incorporating significant amounts of genetic diversity into the expanding human population while simultaneously avoiding the biological consequences of inbreeding. In fact, based upon the genetic evidence that all non-African humans carry small portions of Neanderthal DNA, it appears that a similar type of bestiality did indeed occur with early humans.[44]

Positing one original pair whose offspring then interbred with biological humans is a theory that is not unique to Kemp. The Protestant biologist S. Joshua Swamidass put forward a similar theory recently, but one that places the first theologically human couple only ten thousand years in the past.[45] To do this, Swamidass makes the distinction between genealogical ancestors and genetic ancestors. A genealogical ancestor is someone on our family tree, while genetic ancestors are those with whom we share our genetic history. As Swamidass points out in his book, many of our genealogical ancestors do not pass on genetic information to us: "Genetic ancestry continues to dilute each generation: 1/8, 1/16, 1/32 . . . to a number so small it is unlikely a descendant

42. *CCC* 410.

43. Some have posited that God created Eve with a genetically diverse set of oocytes to increase genetic diversity. Unless she was able to have a few hundred children, this would not help explain the current genetic diversity seen in our present-day population.

44. Villanea and Schraiber, "Multiple Episodes of Interbreeding between Neanderthal and Modern Humans."

45. Swamidass, *The Genealogical Adam and Eve.*

has any genetic material from most of their ancestors."[46] Genetic ancestors, on the other hand, refer to the those from whom we have inherited our genetic makeup.

Because of these differences, one gets a different answer if one looks to find a common genealogical ancestor of all modern humans than if one looks to examine genetic ancestry. Based upon models, Swamidass argues that all humans currently alive could have shared a genealogical ancestral couple within the last ten thousand years: the biblical Adam and Eve in the Garden of Eden.[47] While this seems possible based upon his work, his scenario raises issues for Catholics; namely, this "first couple" would have existed at a time when the population of humans outside the "garden" displayed clear signs of rational conceptual thought. The other individuals around them, some of whom would be our genetic ancestors, would have to be considered theologically human given the strong evidence that they had the capacity for rational thought, worship, and human culture.

In fact, even Swamidass argues that these other individuals who are outside the garden "were fully human in important ways," but he relegates them to "a new category of fully human people," while never quite defining what this new class would be.[48] However, if they are fully human (that is, made in the image and likeness of God), as the archeological evidence from the time suggests, this scenario seems to be incompatible with Catholic theology. As Kemp points out, "Catholic doctrine, however, holds that *all* [theological] human beings who ever lived have original sin, with only two exceptions: Jesus and Mary. . . . Swamidass, however, though a Protestant, gives us a whole tribe more of such exceptions."[49]

46. Swamidass, 36.

47. This means that all modern humans would have had this couple on their genealogical tree, even though there would be many other genetic ancestors that were in existence at that time as well as before that time.

48. Swamidass, 149.

49. Kenneth Kemp, "Problems of Conceiving Human Origins," *Church Life Journal*, August 10, 2020, https://churchlifejournal.nd.edu/articles/problems-of-conceiving-human-origins/.

In Swamidass' version, all of the humans outside the garden would be fully human but not subject to original sin, which occurred via Adam and Eve in the garden. In fact, all the theological humans that had existed over the prior fifty thousand years before the existence of this specific genealogical couple would not have been subject to original sin. So while Swamidass seems to avoid the issue with Kemp's scenario—the requirement that Adam and Eve's offspring would have had to mate with non-theological humans outside the garden—it is at the cost of positing that generations of theological humans existed before Adam and Eve and lived without the effects of original sin.

HUMAN ORIGINS IN POPULATIONS

Obviously, the scenarios described by Swamidass and Kemp are both speculative, and it is important to remember that Genesis uses "figurative language"[50] to describe events that occurred at the dawn of humanity, and that this language is not always clear. For example, it is not obvious what to make of the land of Nod where Cain settled or of the city he built if no other people existed at the time. Likewise, the reference in Genesis 5 to the Nephilim, the warriors of old, who were on the earth at the time is obscure. The point is that understanding this figurative language is not straightforward for reasons discussed previously.[51]

Given the complexities involved, it should not be surprising then that theologians have taken a number of different approaches to the problem modern human genetic diversity poses for a relatively recent monogenic origin of humans. One common approach is to posit that theological humans arose as an initial population rather than as just a single couple. While the *Catechism* does refer to the sin of our "first parents," the ITC document seems to suggest an initial human population as

50. *CCC* 390.

51. See chapter 4 for more on the issues associated with the proper exegesis of Genesis.

a possibility. As Pope Pius XII rightly pointed out in *Humani Generis*, such a polygenic origin of theological humans poses significant questions in relation to the nature and transmission of original sin. The Council of Trent declared that Adam's sin is transmitted to all "by propagation, not by imitation."[52] This means that original sin does not merely refer to people imitating Adam's bad example or imitating Adam's descendants' bad examples. Rather, something is interiorly problematic with us because we have inherited human nature without the accompanying preternatural gifts. Yet, if theological humans emerged in a state of original holiness as a population, the effects of the first sin could not have been propagated to the other members of the group via "propagation." How then could Catholics possibly reconcile such a scenario?

In order to see how this might be possible, it is necessary to step back and examine the transmission of sin in the Adam and Eve account. While both Adam and Eve participated in the original sin of partaking of the fruit of the tree, they each individually made a choice to disobey God's command. Eve rejects God's command first; Adam, through the persuasion of Eve, then follows her in violating the command. Of course, one does not want to read the story too literally, but any scenario of an original couple has this problem. If both parents are to be subjected to the loss of grace, both would have to individually reject God. At the stage of an initial human couple, the effects of original sin are not being transmitted by propagation from one parent to the other but are being propagated by individual acts. Both parents consent to violate the command of God in what could be termed a communal act to which each gives full consent. As a result, there is a communal consent by both parties to reject and fundamentally alter their relationship with God.

52. General Council of Trent, Fifth Session, Decree on Original Sin, June 17, 1546, in *The Sources of Catholic Dogma*, ed. Henry Denzinger and Karl Rahner, trans. R.J. Deferrari (St. Louis, MO: Herder, 1954), 247.

Given this fact, it may be possible that sin (and its effects on human nature) could affect an initial human population in an analogous manner. Just as Adam and Eve consented to an evil act and the rejection of God, it may be possible for a community to consent to an evil act such that each individual bears the consequences by his or her consent to the communal act. In such a scenario, all members of the group would have to consent individually to the act so that all members would be subject to the loss of grace. After such a communal act by this initial population, the effects of sin, a human nature deprived of original justice and holiness, would be transmitted via propagation down through the ages to all of humanity.

Along these lines, it is interesting to note that original sin is traditionally referred to as the sin of Adam rather than the sin of Eve or even the sin of Adam and Eve.[53] Why is this when, in the Genesis account, it is Eve who first violates God's command? The reason the tradition often refers to the sin of Adam (as opposed to the sin of Eve) largely flows from the scriptural understanding of the man serving as the head of the marriage or family. As the head of the marriage, Adam is viewed as in some way responsible for the sin that takes place within the marriage. For example, some commentators have stated that Adam is guilty of the first sin (of omission) by failing to protect his wife from the temptations of the devil.

In an analogous manner, some commentators have viewed Adam as head of an initial population, and as such Adam could be seen as responsible for the introduction of sin into that population.[54] Like a priest is responsible for the salvation of his people, Adam as head of a population would be responsible for

53. Some scriptural passages refer to Eve's responsibility, such a 1 Tim. 2:14, but there is much in the tradition about Adam's ultimate culpability.

54. C. John Collins, "Adam as Federal Head of Humankind," in *Finding Ourselves after Darwin: Conversations on the Image of God, Original Sin, and the Problem of Evil*, ed. Stanley Rosenberg (Ada, MI: Baker Academic, 2018), 143–159.

maintaining them in perfect relation with God. In this scenario, one could condemn Adam for not protecting the community he heads.

Instead of viewing the sin of Adam as referencing his responsibility for the sins of the first human couple, it seems possible that the sin of Adam could refer to his guilt for the individual sins of the first human population. Regardless of whether Adam is the head of a marriage (Adam and Eve) or the head of a community, the initial stages of the transmission of sin through the first population (either a couple or a larger community) do not seem to be addressed specifically by the Council of Trent, nor do they fit neatly with the "by propagation, not by imitation" phrase in the council documents.[55]

When looking at the documents of the council, it is important to remember that it was held at a particular point in history to address particular questions of relevance at the time. This is not to claim the conclusions are relative, but rather to put the statements of the council in the context of the time so that they do not take on a character that was not intended by the council. At the time, the Council of Trent was addressing those who denied the reality of original sin or who thought that we only experience original sin when we fall into sin by imitating the sin of others. To address this issue, the council sought to affirm that all humans have inherited the effects of original sin. Whether this was inherited from a first couple as represented by Adam or from a first population as represented by Adam 1) does not seem to undermine this claim of the council and 2) was not a theological point the council was attempting to address. While

55. The council documents state that "this sin of Adam, which is one in origin and transmitted to all is in each one as his own by propagation, not by imitation." Once this sin occurs, the council indicates that it is inherited via propagation, which is assumed to mean that we inherit this state of human nature deprived of the preternatural gifts from our parents. The origin of this "one" sin—whether it must be the communal act of a couple and not the communal act of a small population—is not addressed by the document. General Council of Trent, Fifth Session, Decree on Original Sin, June 17, 1546, in *The Sources of Catholic Dogma*, ed. Henry Denzinger and Karl Rahner, trans. R.J. Deferrari (St. Louis, MO: Herder, 1954), 247.

it appears that the scenarios outlined above are consistent with the Church's teaching on original sin, this is ultimately an issue that can only be definitively resolved by the Magisterium.

Given that "the transmission of original sin is a mystery that we cannot fully understand,"[56] scenarios regarding its transmission at the dawn of human history will likely always remain inherently speculative. As such, one must remain humble when putting forth scenarios such as Kemp's interbreeding hypothesis or the idea that Adam represented an original population. One must recognize that all these speculative scenarios are open to critique, rejection, or refinement based on new insights, novel scientific data, and further clarifications by the Magisterium.

ORIGINAL SIN AND THE DEVELOPMENT OF DOCTRINE

St. John Henry Newman, in his seminal work *An Essay on the Development of Christian Doctrine*, delineated how our understanding of the truths of the faith can grow over time. According to Newman, doctrinal development is the result not of some break or disruption but of growth over time, a richness that comes to the doctrine through continual engagement and refinement. The Church's understanding of the doctrine of original sin as it relates to the death of unbaptized infants is an excellent example of this type of development. Traditionally, the teaching on unbaptized infants focused on the theory of limbo, an eternal state in which unbaptized infants do not experience the beatific vision (because they died still subject to original sin) but also do not experience any punishment (because they did not commit any personal sin). As a recent Vatican document released by the International Theological Commission points out, "This theory, elaborated by theologians beginning in the Middle Ages, never entered into the dogmatic definitions of

56. *CCC* 404.

the Magisterium, even if that same Magisterium did at times mention the theory in its ordinary teaching up until the Second Vatican Council."[57]

The document then goes on to note that the current *Catechism* makes no mention of limbo: "Rather, the *Catechism* teaches that infants who die without baptism are entrusted by the Church to the mercy of God, as is shown in the specific funeral rite for such children. The principle that God desires the salvation of all people gives rise to the hope that there is a path to salvation for infants who die without baptism (*CCC* 1261)."[58] In its efforts to reconcile its traditional teaching on the necessity of Baptism for salvation with the salvific will of God, the Church's understanding of the fate of unbaptized infants in relation to original sin has developed over time.

According to Newman, such development is something that enriches the faith and allows it to engage and evangelize the world more effectively. As St. John Henry Newman stated, "But whatever be the risk of corruption from intercourse with the world around, such a risk must be encountered if a great idea is duly to be understood, and much more if it is to be fully exhibited. It is elicited and expanded by trial, and battles into perfection and supremacy."[59]

Thus, one has two options when it comes to our current understanding of original sin. The first is to take the doctrine and protect it from any engagement with evolutionary theory. While this might preserve the doctrine from corruption, it does so at the risk of stagnation. The second option is to risk engagement. If the doctrine of original sin is to be duly understood and "fully exhibited," using Newman's terms, it must risk the encounter with new human knowledge. As iron sharpens iron, what is learned *properly*

57. International Theological Commission, "The Hope of Salvation for Infants Who Die Without Being Baptized," preface, 2007, vatican.va.

58. "The Hope of Salvation," preface.

59. John Henry Newman, *An Essay on the Development of Christian Doctrine* (Park Ridge, IL: Word on Fire Classics, 2017), 32.

via science regarding evolution can facilitate developments in our understanding of the doctrine of original sin. As Newman writes, "Doctrines expand variously according to the mind, individual or social, into which they are received; and the peculiarities of the recipient are the regulating power, the law, the organization, or, as it may be called, the form of the development."[60]

The risk, though, is that the doctrine will be corrupted if we merely try to shape it to the new knowledge of the age. If one bends the notion of original sin toward modernist tendencies, one is soon left with the concept that original sin is nothing more than the disordered animal tendencies we inherited at the dawn of our existence from our hominin ancestors. In such a scenario, humankind was flawed from the very beginning of our existence. Yet one can avoid such tendencies if one focuses on maintaining the continuity of the underlying principles regarding original sin, particularly the principles outlined earlier in this chapter. This is not an easy task, but that does not mean one should abandon such efforts, particularly in light of the fact that such efforts can make the doctrine, in the words of Newman, "a living idea." As he states, "a power of development is a proof of life" for a doctrine.[61] It keeps the doctrine from shattering under the weight of new human knowledge while simultaneously allowing it to expand with the addition of new insights.

SUMMARY

Reconciling the data regarding human evolution with the Catholic doctrine of original sin is clearly not a simple task. In fact, it may seem to some that no satisfactory solution is forthcoming. However, rather than assuming the task is insurmountable, one would do well to take counsel from the words that Pope Pius XII wrote in the encyclical *Divino Afflante Spiritu*. While he wrote

60. Newman, 146.

61. Newman, 153.

these words in reference to the difficulties in properly interpreting Scripture, his wisdom applies equally well to the issue of reconciling our scriptural understanding of original sin with human evolution:

> No one will be surprised, if all difficulties are not yet solved and overcome; but that even today serious problems greatly exercise the minds of Catholic exegetes. We should not lose courage on this account; nor should we forget that in the human sciences the same happens as in the natural world; that is to say, new beginnings grow little by little and fruits are gathered only after many labors. . . . Hence there are grounds for hope that those also will by constant effort be at last made clear, which now seem most complicated and difficult.
>
> And if the wished-for solution be slow in coming or does not satisfy us, since perhaps a successful conclusion may be reserved to posterity, let us not wax impatient thereat, seeing that in us also is rightly verified what the Fathers, and especially Augustine, observed in their time viz: God wished difficulties to be scattered through the Sacred Books inspired by Him, in order that we might be urged to read and scrutinize them more intently, and, experiencing in a salutary manner our own limitations, we might be exercised in due submission of mind.[62]

Maintaining humility in the face of this task is essential given the difficulties and complexities involved. While theological speculation on exactly how the first humans appeared and subsequently turned from God can spur interesting and robust discussions, any scenario one can construct, including the ones discussed in this chapter, will always remain speculative and subject to further clarification from the Magisterium. In fact,

62. Pius XII, *Divino Afflante Spiritu* 44–45, encyclical letter, September 30, 1943, vatican.va.

given that these events lie beyond the veil of recorded history, we will likely never know exactly how the first theological humans emerged on our planet or how they broke their relationship with the Creator. As Pope Benedict XVI stated, "It is not the use of weapons or fire, not new methods of cruelty or of useful activity, that constitute man, but rather his ability to be immediately in relation to God. This holds fast to the doctrine of the special creation of man; herein lies the center of belief in creation in the first place. Herein lies the reason why the moment of anthropogenesis cannot possibly be determined by paleontology: anthropogenesis is the rise of the spirit, which cannot be excavated with a shovel."[63]

While the archeological record can give us some guidance on the possible timing of events surrounding human origins, the exact details will always remain speculative, particularly regarding the nature of the fall. The situation is summed up nicely by Fr. Andrew Pinsent:

> Humans, when they awoke to the capacity for abstract thought and moral decisions (however this happened), received also the gift of grace and various other divine gifts to know and love God, as stewards of creation. Yet they freely chose to reject the love of God and so lost these gifts, their nature being stripped bare of grace. Their descendants, who inherit human nature without these gifts, suffer the effects, mostly evident in a bent toward what is broadly acknowledged as evil. Whether or not one chooses to accept this account on moral, philosophical, or theological grounds, what can at least be said is that we are not compelled to reject it on scientific grounds.[64]

63. Benedict XVI, in *Creation and Evolution: A Conference with Pope Benedict XVI in Castel Gandolfo*, ed. Stephan Horn and Siegfried Wiedenhofer (San Francisco: Ignatius, 2008), 15–16.

64. A. Pinsent, "Augustine, Original Sin, and the Naked Ape," in *Finding Ourselves after Darwin: Conversations on the Image of God, Original Sin, and the Problem of Evil*, ed. Stanley Rosenberg (Ada, MI: Baker Academic, 2018), 142.

The key point to recognize, one that Fr. Pinsent stresses, is that one is not required to reject the principles underlying the Catholic doctrine of original sin based upon the science of human evolution. The science, though, does challenge theologians and scientists alike to think about original sin from new and possibly richer perspectives.

10

In the Beginning Was Reason

Matter from Mind

Can reason really renounce its claim to the priority of what is rational over the irrational, without abandoning itself?[1]

—Joseph Ratzinger

There is a common misperception that exists within the culture at large that the advancements of modern science over the past few centuries have rendered religious beliefs obsolete. This thinking is particularly evident when it comes to the science of evolution. The Christian belief that humans are the crown of creation, made in the image and likeness of God, is thought by many to have been replaced by the realization that our species is nothing more than a chance evolutionary accident inhabiting some unremarkable corner of the universe.

However, to accept such a dichotomous view of creation and evolution is to accept a perspective on science and faith that is, at its foundation, not particularly Catholic. It is to see science and religion as two competing alternatives to understanding rather than as two mutually enriching ways of encountering the one unified truth. While there are other factors involved, the widely held perception of an inherent conflict between science and the Catholic Church continues to be fueled by the long shadow of

1. Joseph Ratzinger, *Truth and Tolerance: Christian Belief and World Religions*, trans. Henry Taylor (San Francisco: Ignatius, 2004), 178–183. Reprinted in *Creation and Evolution: A Conference with Pope Benedict XVI in Castel Gandolfo*, ed. Stephan Horn and Siegfried Wiedenhofer (San Francisco: Ignatius, 2007), 20.

the Galileo affair. In the popular retelling of the tale, archaic churchmen suppressed a scientific advance that they feared would undermine the truths of the faith and demote the importance of mankind. In this simplistic rendition of the story, the ultimate triumph of the heliocentric system is championed as the first step in the process by which modern science dethroned man from his privileged place at the center of creation. Eventually, the rise of evolutionary theory set mankind back further, reducing him to a chance by-product of a process that had no intention of producing him.

Yet these supposed "demotions" of man by modern science are more accurately viewed as changes in perspective rather than demotions. For example, in the case of Galileo and the heliocentric system, the Catholic position that man is the crown of creation was in no manner tied to his supposed existence at the "center" of the universe. In fact, in the geocentric Ptolemaic system, man found himself stuck within the lower, corruptible spheres separated from the privileged, incorruptible, perfect spheres of the heavens. Within the geocentric system prevalent during the sixteenth century, mankind found himself in the unenviable position of being just above the pit of hell.

In addition, the eventual acceptance of the heliocentric model did not undermine Church teaching despite the fears of those who opposed Galileo at the time. Rather, the rigorous scientific debates regarding the solar system and the heavens that played out in the sixteenth and seventeenth centuries represented an important step in the ongoing enterprise of scientific discovery, an enterprise that continues to reveal, in quite profound ways, the majesty of God's creation. It is a journey that led us from a view of the universe in which the Earth was fixed in a low and corruptible point near the center to one in which the Earth is a rare hospitable planet in a vast and ancient universe. While such a view required a more sophisticated exegesis of certain biblical texts (a sticking point in the Galileo affair), it did not overthrow or displace any of the central tenets or dogmas of the Catholic faith. The reality of the Incarnation, or the Real Presence of the Eucharist, or the nature of man, or the goodness of creation did not lose force when we migrated from the old, static geocentric cosmology to the modern view of an expansive fourteen-billion-year-old universe. In fact, such a transition opened new and exciting opportunities for theological reflection on the nature of such an evolving creation.

In a similar manner, the data of evolutionary theory did not necessarily demote the importance of man, nor should it put Catholics on the defensive against the so-called encroachments of modern science. As St. John Henry Newman stated, "Mr. Darwin's theory *need* not, then, be atheistical, be it true or not; it may simply be suggesting a larger idea of Divine Prescience and Skill."[2]

As this great saint and scholar recognized, the science of evolution does not undermine the Catholic understanding of what man is—a unity of body and soul made in the image and likeness of God. However, it does require unique reflections on the manner by which, in line with God's providence, man came to be. Newman recognized this, pointing out that certain ways of interpreting Genesis would not be compatible with evolution. However, as he reflected on the meaning of the Genesis text, he was clear that evolution did not lead to any contradictions with the faith: "I don't know why Adam needs be immediately out of dust—*Formavit Deus hominem de limo terrae* ["God formed man from the dust of the ground" (Gen. 2:7)]—i.e. out of what really was dust and mud in nature, before He made it what it was, living."[3] For Newman, whether man was brought into being straight from the dust of the ground or his origins were mediated

2. John Henry Newman, "Letter to J. Walker of Scarborough on Darwin's Theory of Evolution," May 22, 1868, https://inters.org/Newman-Scarborough-Darwin-Evolution.

3. John Henry Newman, "Letter to Edward Bouverie Pusey," June 5, 1870, in *Letters and Diaries of John Henry Newman*, vol. 25, *The Vatican Council, January 1870–December 1871*, ed. Charles Stephen Dessain (Nashville, TN: Thomas Nelson, 1961), 138.

through physical material provided by an advanced hominin, man was still the outcome of what he referred to as Divine Prescience and Skill.

THE TWO BOOKS

Much of the perceived conflict between evolution and Catholicism stems from a failure to distinguish what can legitimately be read from the Book of Nature and what can legitimately be read from the Book of Scripture. While scientific discovery—the "reading" of the Book of Nature—can uncover how the planets move or how species are related, it cannot fully explain the purpose of man or answer why a universe that is ordered to support life exists in the first place.

This is in no way meant to diminish the importance and success of modern science. In fact, scientists, through Herculean efforts of human ingenuity and dedication, have been able to offer compelling descriptions of the structure and order of the universe from the cosmic-galactic scale down to the subatomic scale. While these descriptions are subject to refinement and change, they have provided us with a solid foundation of knowledge regarding the physical world. Despite these successes, scientists, no matter how many experiments they perform and no matter how many ingenious theories they develop, can never develop scientific explanations for *why* we happen to exist in such an ordered universe. They can describe the order of the universe at all different scales, but they cannot explain its *ultimate* source.

While science cannot answer these questions, it can point us in the proper direction. In fact, the success of science, the fact that our human mind has been capable of discovering the laws and order of nature, is, in and of itself, reason to suspect something more lies beyond the physical world. For a nonbeliever, the fact that human reason is capable of uncovering and discerning the fabric and structure of creation can only be attributed to a cosmic accident. For believers, though, this curious coincidence points to a Creative Reason, an Author of creation who speaks to us through it. As the *Catechism* states, "Our human understanding, which shares in the light of the divine intellect, can understand what God tells us by means of his creation."[4]

From a Catholic perspective then, it should not be surprising that when one reflects upon the knowledge gained from reading the Book of Nature, one is often drawn toward the ultimate source of reason that stands behind it. It is why the book of Wisdom states, "For from the greatness and beauty of created things comes a corresponding perception of their Creator" (Wis. 13:5). It is why Einstein, though not himself a believer, could say, "My religiosity consists in a humble admiration of the infinitely superior spirit that reveals itself in the little that we, with our weak and transitory understanding, can comprehend of reality."[5] Even committed atheists speak in reverent tones regarding the Book of Nature, for it is hard for anyone who examines it to escape unscathed by wonder.

For some, this can become the first step toward belief, as the wonder of scientific discovery can lead one toward the source of wonder. However, it is important to recognize that scientific discovery, regardless of how beautiful or profound it is, does not allow one to develop a mathematical proof for God's existence or to develop a scientific theory of a Creator. That is not the role of science. The discoveries of science can bring one to contemplate the ultimate questions of meaning and purpose, but they cannot on their own answer them. For that, one must turn to philosophy and theology.

This is where the Book of Scripture provides guidance, as it speaks to these questions of ultimate meaning and purpose. It is within Scripture that God shines light on these questions, ones

4. *Catechism of the Catholic Church* 299.

5. Albert Einstein, in *Albert Einstein, The Human Side: Glimpses from His Archives*, ed. Helen Dukas and Banesh Hoffmann (Princeton: Princeton University Press, 2013), 66.

that science cannot answer no matter how deeply it scrutinizes its subject matter. Scripture, while it does not explicate the physical processes undergirding the universe, can reveal the depth and meaning of both man and creation. As the *Catechism* states when discussing the Genesis text, "It is not a question of knowing when and how the universe arose physically, or when man appeared, but rather of discovering the meaning of such an origin."[6] In the words of Joseph Ratzinger, "[The Genesis text] explains [our] inmost origin and casts light on the project that [we] are."[7]

If Scripture is read in this manner, the two books can become complementary. However, this only works if we take *both* books seriously. To understand man, it is necessary to study his biology and his physical ties to other primates as described via the science of evolution. However, one must not lose sight of the fact that man is a unity of body and spirit, created in the *imago dei* with a heavenly destiny as revealed through Scripture. Together these two books, if read correctly, can generate an integrated picture of man.

However, if one attempts to use the Book of Nature to describe man's ultimate purpose or the Book of Scripture to explain the material origin of man's body, one is bound to produce conflict. In fact, nearly all of the apparent conflict between evolution and the faith discussed in these pages has at its root the abuse of one of these two great books. If used properly, though, each within its own sphere with its proper methods, what the science of evolution tells us about man will not ultimately conflict with what the Book of Scripture reveals regarding what man is. Of course, integrating these two books is not always straightforward, as the Galileo episode so aptly illustrates. Yet, despite such difficulties, Catholics have no reason to cower in fear when it comes to the evolution/creation issue, armed as we are with the confidence

6. *CCC* 284.

7. Joseph Ratzinger, *"In the Beginning . . .": A Catholic Understanding of the Story of Creation and the Fall* (Grand Rapids, MI: Eerdmans, 1995), 50.

that truth cannot contradict truth. Rather, Catholics should view the unresolved questions at the heart of the evolution/creation discussion as unique opportunities to come to a deeper understanding of the mystery of man, a physical being embedded with the instincts and physiology of our primate ancestors yet also a spiritual being created to know and love God in eternity.

EVOLUTION AND PURPOSE

Arguably *the* central problem in the evolution/creation debate is the fact that many people erroneously read the Book of Nature in a manner that is incompatible with the Catholic understanding that creation has a purpose. Unfortunately, as science advances and is able to describe more and more of the evolutionary history of the cosmos, many people read this as evidence that the universe is, at heart, devoid of purpose and meaning. If we can invoke cold scientific explanations for this evolutionary story, both cosmically and biologically, it seems that there is no room left in the picture for God or his purposes. This is exactly the message the evolutionary biologist George Gaylord Simpson read from the Book of Nature, claiming, "Man is the result of a purposeless and natural process that did not have him in mind. He was not planned. He is a state of matter."[8]

Simpson is not alone in reading this message. Many people, both atheists and Christians alike, tend to see in evolution a process that has no direction or purpose. The prevalence of this view in the general populace should come as no surprise given that nearly every scientific textbook on evolution dogmatically insists evolution is directionless, that it has no end goal. Yet the actual evidence of evolution suggests otherwise.

It is true that when one focuses in on a specific narrow branch of evolution, the process may seem to meander aimlessly. In fact,

8. G.G. Simpson, *The Meaning of Evolution* (New Haven, CT: Yale University Press, 1967), 344.

in some circumstances, such as cave animals that lose their sense of sight or tetrapods that lose their limbs, a specific local evolutionary lineage can move toward less complexity. However, when one steps back and views the entire grand sweep of biological evolution, one can clearly see a movement toward increased diversity and complexity in the biosphere. There appears to be a directionality toward greater complexity baked into evolution as a whole.

As the cognitive neuroscientist, science writer, and nontheist Bobby Azarian points out, "Unambiguous evidence of continued complexification over time can be found in the same fossil record that confirmed Darwin's great theory."[9] If one looks back at the Earth 2.5 billion years ago, the only organisms were relatively small prokaryotic cells that seemingly could not produce enough energy to develop into larger, more complex forms. But these cells provided the foundation for the next revolution in complexity, the eukaryotic cell. In fact, cooperative interactions between small prokaryotic cells facilitated the emergence of eukaryotic cells, which could reach sizes and organizational complexities that were orders of magnitude higher than prokaryotic cells.[10]

These individual eukaryotic cells then laid the foundation for the next revolution, again through cooperative interactions. They began to interact with groups of other cells and form cooperative units that we recognize now as multicellular organisms. This increase in cooperation among eukaryotic cells allowed for the emergence of a whole new level of complexity. As more specialization and cooperation occurred among the cells that composed these organisms, novel complexities such as nervous systems, skeletons, and muscles emerged to allow motile sentient creatures to roam the planet.

Such cooperation-driven increases in complexity *are* the story

9. Bobby Azarian, *The Romance of Reality* (Dallas: BenBella Books, 2022), 120.

10. The evidence indicates that it was the process of endosymbiosis, in which one primitive cell took up residence inside another primitive cell, that led to the establishment of the more complex eukaryotic cells.

of evolution on this planet. Based upon the scientific evidence, it is hard to deny that the grand scope of evolution, as it has actually occurred on this planet, is one that reveals a directional increase in complexity. It is the story of the fossil record. Even nontheists such as the biologist E.O. Wilson agree on this point: "Many reversals have occurred along the way, but the overall average across the history of life has moved from the simple and few to the more complex and numerous. During the past billion years, animals as a whole evolved upward in body size, feeding and defensive techniques, brain and behavioral complexity, social organization, and precision of environmental control. . . . More precisely the overall average of these traits and their upper extremes went up."[11]

What then is the underlying cause of the cooperation-driven increase in complexity seen in evolution? Is there a scientific explanation for this? There could be. As Azarian points out, "You can have inevitable progress without any supernatural or conscious force guiding or driving the evolutionary process."[12] The entire system could be set up for progress such that no direct supernatural intervention is necessary.

In the case of our cosmos, it seems that the information contained in the highly ordered foundational laws and forces of the universe (discussed in more detail in chapter 6) could, over billions of years, drive the organization of matter into certain elements in specific proportions, like large amounts of hydrogen and carbon and low amounts of manganese. The chemical information contained in these elements then could drive them to combine in specified ways based on chemical rules to produce molecules such as amino acids, formaldehyde, and water, molecules that just happen to be found throughout the cosmos. The information in these simple molecules then allows the possibility of them combining in specific ways to make life possible.

11. E.O. Wilson, *The Diversity of Life* (Cambridge, MA: Harvard University Press, 1992), 187.

12. Azarian, *The Romance of Reality*, 115.

Once an ecosystem of life emerges with its higher ability to store information, as Azarian suggests, this "self-correcting system [of living organisms] grows more robust and computationally powerful because it is always solving survival problems and storing solutions in memory."[13] If organisms are to survive, they must continually adapt by building upon and sharing these solutions in cooperative ways that generate more complexity and more diverse forms over time.

It seems that there may be something about an information-laden universe, the type we find ourselves in, that can explain scientifically the drive toward higher complexity that we see in the evolutionary process. But such an explanation, however tentative, doesn't really answer the foundational human question: Why do we find ourselves in a universe that seems inherently driven toward more complexity? Science may in theory be able to explain *how* the universe is set up to evolve toward increasing complexity, but it can't explain *why* it has been set up that way. The question of what it all means remains unresolved by science.

SALVATION HISTORY AND EVOLUTIONARY HISTORY

If evolution is moving toward more complexity, if it is a progressive story that is unfolding in time, what is the ultimate meaning of this story? Before investigating this, it is important to recognize that the question of the meaning of evolution cannot be resolved within the science itself. Modern science, in order to gain precision and detail, has deliberately cut itself off from directly tackling the deeper questions of ultimate purpose and meaning. However, one cannot dismiss the question simply because it goes beyond the narrow confines of modern science. It remains a real question with a real answer. Either evolution has meaning or it doesn't. Pope Benedict XVI is clear on this point: "Of course the question remains open whether . . . evolution as a whole—has a meaning,

13. Azarian, 115.

[but] it cannot be decided within the theory of evolution itself; for that theory this is a methodologically foreign question, although of course for a live human being it is the fundamental question of the whole thing."[14]

How then do we answer this? What is the key through which we can interpret the meaning of evolution? Many atheists and agnostics use scientism as their interpretive key, arguing that meaning and purpose must be absent from evolution because things like meaning and purpose don't really exist. But this is akin to strapping on a blindfold and then assuming, because one does not see it, that light does not exist.

For Catholics, though, we are blessed to have a much more robust interpretative key through which to view reality: the lens of salvation history. While the biologist Theodosius Dobzhansky famously stated that "nothing in biology makes sense except in the light of evolution," as Catholics, we could rephrase this to say that nothing in the cosmos makes sense except in the light of salvation history.

In viewing evolution in light of salvation history, it is hard not to be struck by the many parallels between the two stories. To begin with, evolution and salvation history are both stories that are progressing toward an end. Salvation history is progressing toward Christ and is hopefully moving toward our ultimate sanctification and deification in and through him. Evolution demonstrates that the universe is progressing toward something as well, that evolution has an inherent direction. In fact, the *Catechism* resonates with this fact, stating that "the universe was created 'in a state of journeying' toward an ultimate perfection yet to be attained, to which God has destined it."[15] Is not evolution progressing toward the origin of that complex being that can know, love, and serve Christ for all eternity? Is it not providing the

14. Joseph Ratzinger, "Schöpfungsglaube and Evolutionstheorie," in *Wer ist das eigentlich—Gott?* ed. H.J. Schulz (Munich: Kösel, 1969). Reprinted in *Creation and Evolution*, 12.

15. *CCC* 302.

creature that makes the story of salvation history necessary? The two stories seem to converge in the human person, the creature that ultimately links the two together. But should we not expect this convergence if the same God is the author of both stories?

This convergence is even more striking if one examines how both stories prominently feature creaturely freedom within their plots. In the evolutionary story, creaturely causation plays an essential role. As discussed in chapter 5, chance encounters of creatures can shape the evolutionary trajectory, as can the response of organisms to their environment. Evolution reveals a created world in which God does not control everything like a despot or a tyrant. Rather, he allows creatures to play an active role in his story of life. One might object that creatures acting on their own accord could thwart his plan for evolution. But as the theologian Matthew Ramage points out, God's providence cannot be undone by creaturely actions:

> Creation does not merely follow the preprogrammed instructions of a divine craftsman, nor does God "tweak" his story or score in the middle of its performance as envisioned by some Christians. On the contrary, evolutionary adaptations arise naturally from within the story of creation itself, according to the rules of its nature instilled therein by its divine author. But why would this author allow his script to seemingly go off the rails at points? In brief, it is because God has bestowed on creatures the dignity of being real causes. And their errors—our errors—are all known to God and allowed as part of his plan.[16]

In a similar manner, salvation history is not rigidly scripted and controlled by the Creator. Rather, it is radically open to the participation of humans, who do not know his plans and seemingly send the plan "off the rails at points." But despite the

16. Matthew Ramage, *From the Dust of the Earth: Benedict XVI, the Bible, and the Theory of Evolution* (Washington, DC: The Catholic University of America Press, 2022), 75.

foibles of David or the failings of Moses or the idol worshiping of Israel, God's plan for salvation history is not thwarted by human freedom but is all the richer for it. In both evolutionary history and salvation history, God reveals his power by achieving his ends *with* and *through* the inherent limitations of his creatures.

Another central theme of evolution is that of death and rebirth. Too often theists recoil from the death, destruction, and extinction that is endemic to the evolutionary process. How could a loving God create in such a fashion? Yet, throughout the evolutionary process, death and extinction do not have the final say. The major extinction events in evolutionary history open the possibility for rebirth. Every extinction event is followed by the emergence of novel complex life forms that make possible the next stage of evolution. For example, the extinction at the end of the Ediacaran was followed by the explosion of novel complex animal forms in the Cambrian. A similar situation occurred at the end of the Permian as well as at the end of the Cretaceous. In each instance of mass extinction, death does not have the final say.

This theme also redounds in salvation history. The destruction of the first temple in Jerusalem and the Babylonian exile seemed to be the end of the Jewish people, but from this "death" the second temple emerged, one arguably more central to Jewish life than the first. Likewise, the blood and death of the early Christian martyrs, which on worldly terms did not seem to bode well for the long-term health of the Church, laid the foundation for the rise of Christendom. In each of these cases, just as in evolution, death did not have the final say. Of course, this is because salvation history has as its central event the Crucifixion, Death, and Resurrection of Christ. Everything, all of salvation history and all of evolutionary history, needs to be viewed through this pivotal event.

It is interesting that many of the same issues that give Catholics pause for concern about evolution, like the role of creaturely causation and the role of death and destruction, are deeply embedded within the story of salvation history. But should this be surprising? Why should the story of evolution not reflect the story of salvation history, with its plot twists and turns and its intimate involvement of creatures who play a role in God's plan, often unwittingly? Why should it not reflect the central theme of salvation history, that out of death comes new life? Of course, this claim goes well beyond the science, but so do all claims that attempt to extract meaning (or lack of meaning) from evolution. This does not make them invalid; it merely means they are built upon a wider wisdom than science.

TOWARD A THEOLOGY OF EVOLUTION

John Paul II made a concerted effort throughout his pontificate to foster a fruitful dialogue between the Church and modern science. He respected the autonomy of science, recognizing that it had its "own principles, its pattern of procedures, its diversities of interpretation and its own conclusions."[17] However, he also recognized that the scientific enterprise was not fully autonomous. He sensed the danger that occurs when branches of knowledge become siloed and walled off from each other, a phenomenon all too common in the modern university. When distinct branches of knowledge become disengaged with one another, they tend to overstep their bounds and transgress into areas beyond their competency. This is particularly true of science and theology, as John Paul II explains: "Only a dynamic relationship between theology and science can reveal those limits which support the integrity of either discipline, so that theology does not profess a pseudo-science and science does not become an unconscious theology. . . . The uses of science have on more than one occasion proved massively destructive, and the reflections on religion have

17. John Paul II, "Letter of His Holiness John Paul II to Reverend George V. Coyne, SJ, Director of the Vatican Observatory," June 1, 1988, vatican.va.

too often been sterile. We need each other to be what we must be, what we are called to be."[18]

To reach their potential in uncovering the truth, science and religion both need the stabilizing influence of the other. If science cuts itself off from any discussion of ultimate meaning and value, it runs the risk of becoming a dehumanizing force and being "massively destructive." If religion cuts itself off from science, it risks making a mockery of the divine gift of human reason. As John Paul II stated, "Science can purify religion from error and superstition. Religion can purify science from idolatry and false absolutes. Each can draw the other into a wider world, a world in which both can flourish."[19]

This is a bold task, one that can only be done when each discipline takes the other seriously and looks upon its own conclusions with humility. The importance of this is nowhere more evident than in the evolution/creation discussion as it relates to the nature of man. Within the Catholic tradition, mankind has been viewed as the crown of creation, not because he was created at the end of the six days described in the Genesis text, but rather because he is the only earthly creature with the capacity to know and love God. He is the only earthly creature for which God suffered and died. A scenario in which man's physical form emerged through an evolutionary process in no manner undermines this theological reality; rather, it has the potential to shed new light on it. In a similar manner, the theological understanding of man as made in the *imago dei* can shed light on our understanding of evolution and the evolutionary process.

Our theological understanding does this not by undermining the truths of science, but rather by illuminating them. As C.S. Lewis famously stated, "I believe in Christianity as I believe that the Sun has risen, not only because I see it, but because by it, I

18. John Paul II, "Letter to Reverend George V. Coyne."

19. John Paul II, "Letter to Reverend George V. Coyne."

see everything else."[20] This is the role that theology plays in the understanding of evolution. It allows us to see evolution in the proper scope and context, to understand its explanatory limits, and to shine light on its true meaning. This is why St. Anselm stated, "I do not seek to understand in order that I may believe, but rather, I believe in order that I may understand."[21]

For those who try to uncover the "meaning" of evolution without any recourse to a belief in a Creator God, they are bound to find it devoid of meaning. In fact, their starting assumption necessitates their conclusion, not the scientific data. If one assumes that the universe 1) simply exists as a brute fact and 2) is nothing but atoms colliding in a void, there is simply no space for meaning and purpose. Their materialistic, atheistic axioms dictate this conclusion.

To be fair, the same could be said of a theistic reading of evolution. If one starts an examination of the science with a belief that the Creator endowed creation with order, purpose, and meaning, one is necessarily going to find purpose and meaning in all of creation. One's theistic axioms dictate this conclusion.

If our first principles then dictate our conclusions regarding purpose and evolution, how does one discern which of these two alternatives is more fitting? Is either starting point, materialism or theism, just as valid? C.S. Lewis, in the quote above, points us toward the answer: one must determine which of the two views of reality best illuminates our understanding of "everything else." It is in this respect that the materialistic starting point is found wanting.

The underlying issue is that a materialistic, atheistic approach to evolution obscures rather than illuminates our human understanding. For starters, such an approach undermines our ability to rely upon our own reason, as it reduces "reason" to a chance

20. C.S. Lewis, "Is Theology Poetry?" in *The Weight of Glory* (San Francisco: HarperOne, 1949), 116–140.

21. Anselm, *Proslogion*, trans. M.J. Charlesworth (Notre Dame, IN: Notre Dame University Press, 1979), chapter 1.

evolutionary by-product that has emerged merely as a survival advantage and therefore has no necessary connection to the truth. Yet all of us, atheists included, live our lives as if we *can* know the truth about things through reason. One cannot be a committed atheist if one does not believe that atheism is correct, that it is a truthful proposition. Likewise, we all act as if there is an ultimate purpose to our lives, and we make moral judgments on our actions and on the actions of those around us. This behavior is ubiquitous among humans even if the moral judgments we make vary from person to person. If evolution is nothing more than a process by which matter aggregates and disaggregates, then these universal human experiences, the convictions that we can know the truth and that we can judge actions as right or wrong, become mere illusions foisted upon us by our evolutionary past. In this scenario, our entire shared experiential reality of being human and living in human societies where we seek understanding and make moral judgments disintegrates into a meaningless jumble.

However, if the wisdom of theology is applied to evolution, nature and creation are revealed in an entirely different light. Evolution is transformed from an inexplicable cosmic coincidence into the project of a loving Creator who has gifted the world with "endless forms most beautiful."[22] The order that is necessary for evolution and allows it to proceed along an emergent process of complexification becomes a reflection of Reason itself rather than a fluke accident. Our uncanny ability to understand the evolutionary process, or anything else for that matter, is not illusory but is the result of humanity being formed by the source of all reason.

With his typical insight and brilliance, Pope Benedict XVI distills the entire creation/evolution discussion down to a single fundamental question: "The question is whether reason, or rationality, stands at the beginning of all things and upon their foundation or not."[23] This is the question that we face, a question

22. Charles Darwin, *Origin of Species* (New York: Mentor, 1958), 459.

23. Ratzinger, *Truth and Tolerance*, 178–183. Reprinted in *Creation and Evolution*, 19.

that has at root only two options. Either creation stands upon the foundation of reason, or it dissolves upon the shifting sands of irrationality. If at the beginning lies nothing more than blind irrational chance, then our human capacity for rational thought renounces all claims upon the truth, and the arguments in this book—and those found in all books for that matter—are for naught. It is only through recognizing that reason is the foundation of all things that the world and everything in it has any hope of being comprehensible.

Yet even if our starting point is correct, integrating evolution and creation is no easy task because our human reason, though a reflection of divine reason, is marred by human frailty and failings. As such, we tend to travel down blind paths and reason our way into apparent contradictions regarding evolution and creation. These apparent contradictions tend to harden into false dichotomies, when instead they should unveil fresh vistas for theological and scientific reflection. If this book has made one thing clear, it is that in the evolution/creation discussion, there are still open and unresolved questions. Rather than being a source of unease and unnecessary conflict, though, these questions and apparent difficulties should be a source of hope; they should be the means through which we can move toward a deeper and richer understanding of reality. From a Catholic perspective, one should have every confidence that by wrestling with such questions, aided by the proper application of human reason enlightened by faith, we will come to a fuller appreciation of both what it means to have evolved from other primates and what it means to be created in God's image.

Bibliography

Adler, Mortimer. *The Difference of Man and the Difference It Makes*. New York: Fordham University Press, 1993.

Alex, Bridget. "*Homo heidelbergensis*: The Answer to a Mysterious Period in Human History?" *Discover Magazine*, September 16, 2009, https://www.discovermagazine.com/planet-earth/homo-heidelbergensis-the-answer-to-a-mysterious-period-in-human-history.

Anselm. *Proslogion*. Translated by M.J. Charlesworth. Notre Dame, IN: Notre Dame University Press, 1979.

Anton, Susan C., et al. "Morphological Variation in *Homo erectus* and the Origins of Developmental Plasticity." *Philosophical Transactions of the Royal Society B* 371 (2016). http://doi.org/10.1098/rstb.2015.0236.

Artigas, Mariano, et al. *Negotiating Darwin: The Vatican Confronts Evolution 1877–1902*. Baltimore, MD: Johns Hopkins University Press, 2006.

Aubert, M., et al. "Earliest Hunting Scene in Prehistoric Art." *Nature* 576 (2019): 442–445. https://doi.org/10.1038/s41586-019-1806-y.

Augustine. *The Literal Meaning of Genesis*. Translated by John Taylor. New York: Newman, 1982.

———. *The Trinity*. Translated by Edmund Hill, OP. Brooklyn: New City, 1991.

Austriaco, Nicanor, et al. *Thomistic Evolution: A Catholic Approach to Understanding Evolution in the Light of Faith*. Providence, RI: Cluny Media, 2019.

Azarian, Bobby. *The Romance of Reality*. Dallas: BenBella Books, 2022.

Baglow, Christopher. *Creation: A Catholic's Guide to God and the Universe*. Notre Dame, IN: Ave Maria, 2021.

Baglow, Christopher. *Faith, Science and Reason: Theology on the Cutting Edge*. 2nd ed. Downers Grove, IL: Midwest Theological Forum, 2019.

Barr, Stephen. *The Believing Scientist: Essays on Science and Religion*. Grand Rapids, MI: Eerdmans, 2016.

———. *Modern Physics and Ancient Faith*. South Bend, IN: University of Notre Dame Press, 2003.

Behe, Michael. *Darwin's Black Box: The Biochemical Challenge to Evolution*. New York: Free Press, 1996.

———. *The Edge of Evolution: The Search for the Limits of Darwinism*. New York: Free Press, 2007.

Benedict XVI. *Creation and Evolution: A Conference with Pope Benedict XVI in Castel Gandolfo*. Edited by Stephan Horn and Siegfried Wiedenhofer. San Francisco: Ignatius, 2007.

———. "Homily of His Holiness Benedict XVI." April 24, 2005, vatican.va.

———. "Pope's Words to Cameroon Muslim Leaders." May 19, 2009, https://www.catholicculture.org/culture/library/view.cfm?recnum=8830.

Bourque, G., et al. "Ten Things You Should Know about Transposable Elements." *Genome Biology* 19, no. 199 (2018). https://doi.org/10.1186/s13059-018-1577-z.

Bradley, James. "Random Numbers and God's Nature." In *Abraham's Dice: Chance and Providence in the Monotheistic Traditions*. Edited by Karl W. Giberson. Oxford: Oxford University Press, 2016.

Callaway, E. "Genetic Adam and Eve Did Not Live Too Far Apart in Time." *Nature* (2013), https://doi.org/10.1038/nature.2013.13478.

———. "Oldest *Homo Sapiens* Fossil Claim Rewrites Our Species' History." *Nature* (2017). https://doi.org/10.1038/nature.2017.22114.

Carl, Brian. "Thomas Aquinas on the Proportionate Causes of Living Species." *Scientia et Fides* 8, no. 2 (2020).

Carroll, Lewis. *Alice in Wonderland*. Orinda, CA: Seawolf, 1871.

Carroll, Sean. *Endless Forms Most Beautiful: The New Science of Evo Devo*. New York: W.W. Norton, 2006.

Carroll, William. "Aquinas and the Big Bang." *First Things* 97 (November 1999): 18–20.

———. "Stephen Hawking's Creation Confusion." *Public Discourse*, September 8, 2010, https://www.thepublicdiscourse.com/2010/09/1571/.

Catechism of the Catholic Church. 2nd ed. Vatican City: Libreria Editrice Vaticana, 1997.

Chiu, Lynn. *Extended Evolutionary Synthesis: A Review of the Latest Scientific Research*. West Conshohocken, PA: John Templeton Foundation, 2022.

Chothia, C. "Principles that Determine the Structure of Proteins." *Annual Review of Biochemistry* 53 (1984): 537–572.

Clack, Jennifer A. "The Fish–Tetrapod Transition: New Fossils and Interpretations." *Evolution: Education and Outreach* 2 (2009): 213–223. https://doi.org/10.1007/s12052-009-0119-2.

Collard, M., and B.A. Wood. "Defining the Genus *Homo*." In *Handbook of Paleoanthropology*, 2nd ed. Edited by W. Henke and I. Tattersall, 2108–2144. Berlin: Springer, 2015.

Collins, C. John. "Adam as Federal Head of Humankind." In *Finding Ourselves after Darwin: Conversations on the Image of God, Original Sin, and the Problem of Evil*. Edited by Stanley Rosenberg, 143–159. Ada, MI: Baker Academic, 2018.

Conard, N.J. "A Critical View of the Evidence for a Southern African Origin of Behavioural Modernity." *South African Archaeological Society Goodwin Series* 10 (2008): 175–179.

Congregation for the Doctrine of the Faith. "Doctrinal Commentary on the Concluding Formula of the *Professio Fidei*." *L'Osservatore Romano*, July 15, 1998, https://www.ewtn.com/catholicism/library/doctrinal-commentary-on-concluding-formula-of-professio-fidei-2038.

Conway Morris, Simon. *Life's Solution: Inevitable Humans in a Lonely Universe*. Cambridge: Cambridge University Press, 2003.

Cropping, Jasper. "Traditional Surnames Are Becoming Extinct: Farewell to the Footheads and Pauncefoots." *The Telegraph*, November 18, 2002, https://www.telegraph.co.uk/news/newstopics/howaboutthat/9685356/Traditional-surnames-are-becoming-extinct-farewell-to-the-Footheads-and-Pauncefoots.html.

Cunningham, Conor. *Darwin's Pious Idea: Why the Ultra-Darwinists and Creationists Both Get It Wrong*. Grand Rapids, MI: Eerdmans, 2010.

Cyran, Krzysztof A., and Marek Kimmel. "Alternatives to the Wright–Fisher Model: The Robustness of Mitochondrial Eve Dating." *Theoretical Population Biology* 78, no. 3 (2010): 165–172. https://doi.org/10.1016/j.tpb.2010.06.001.

d'Errico, F., et al. "Early Evidence of San Material Culture Represented by Organic Artifacts from Border Cave, South Africa." *Proceedings of the National Academy of Sciences of the United States of America* 109, no. 33 (2012): 13214–13219. https://doi.org/10.1073/pnas.1204213109.

Darwin, Charles. *The Descent of Man, and Selection in Relation to Sex*. London: Penguin Classics, 2004.

———. "Letter to Asa Gray," November 26, 1860. Darwin Correspondence Project website, https://www.darwinproject.ac.uk/letter/DCP-LETT-2998.xml.

———. *The Origin of Species*. New York: Mentor, 1958.

Dawkins, Richard. *The Blind Watchmaker*. New York: W.W. Norton, 1986.

———. *River Out of Eden*. New York: Basic Books, 1995.

———. Speech at the Edinburgh International Science Festival, April 15, 1992. Reprinted as "EDITORIAL: A Scientist's Case against God," *The Independent*, 1992.

Dawkins, Richard. "You Can't Have It Both Ways: Irreconcilable Differences." In *Science and Religion: Are They Compatible?* Edited by Paul Kurtz. Buffalo, NY: Prometheus Books, 2003.

Denton, Michael, Peter Dearden, and Stephen Sowerby. "Physical Law Not Natural Selection as the Major Determinant of Biological Complexity in the Subcellular Realm: New Support for the Pre-Darwinian Conception of Evolution by Natural Law." *BioSystems* 71 (2003): 297303.

Dobzhansky, Theodosius. *The Biology of Ultimate Concern*. New York: New American Library, 1967.

Einstein, Albert. *Albert Einstein, The Human Side: Glimpses from His Archives*. Edited by Helen Dukas and Banesh Hoffmann. Princeton: Princeton University Press, 2013.

———. *Mein Weltbild*. Edited by Carl Seelig. Stuttgart-Zurich-Vienna: Europa, 1953.

Fairbanks, Daniel, and Peter Maughan. "Evolution of the *NANOG* pseudogene family in the human and chimpanzee genomes." *BMC Evolutionary Biology* 6, no. 12 (2006), https://doi.org/10.1186/1471-2148-6-12.

Fehlner, Peter D. "In the Beginning." Kolbe Center for the Study of Creation, November 1, 2001, https://www.kolbecenter.org/in-the-beginning/.

Feser, Ed. "Blinded by Scientism." *Public Discourse*, March 9, 2010, https://www.thepublicdiscourse.com/2010/03/1174/.

Futuyma, Douglas. *Science on Trial: The Case for Evolution*. New York: Pantheon Books, 1982.

Galhardo, R.S., P.J. Hastings, and S.M. Rosenberg. "Mutation as a Stress Response and the Regulation of Evolvability." *Critical Reviews in Biochemistry Molecular Biology* 42, no. 5 (2007): 399–435. https://doi.org/10.1080/10409230701648502.

Galileo Galilei. "The First Day." In *Dialogue Concerning the Two Chief World Systems*. 2nd ed. Translated by Stillman Drake. Berkeley: University of California Press, 1967.

Gould, Stephen J. *The Structure of Evolutionary Theory*. Cambridge, MA: Belknap, 2002.

Graney, Christopher. "An Alternative History of Modern Science." *Church Life Journal*, January 14, 2020, https://churchlifejournal.nd.edu/articles/an-alternate-history-of-modern-science/.

Handwerk, Brian. "An Evolutionary Timeline of Homo Sapiens." *Smithsonian Magazine*, February 2, 2021, https://www.smithsonianmag.com/science-nature/essential-timeline-understanding-evolution-homo-sapiens-180976807/.

Hawkins, M.B., et al. "Latent Developmental Potential to Form Limb-like Skeletal Structures in Zebrafish." *Cell* 184, no. 4 (2021): 899–911.

Henn, B.M., et al. "The Great Human Expansion." *PNAS* 109, no. 44 (2012): 17758–17764. www.pnas.org/cgi/doi/10.1073/pnas.1212380109

Hoffmann, D.L., et al. "U-Th Dating of Carbonate Crusts Reveals Neandertal Origin of Iberian Cave Art." *Science* 359 (2018): 912–915. https://doi.org/10.1126/science.aap7778.

Iglesias-Groth, Susana. "A Search for Tryptophan in the Gas of the IC 348 Star Cluster of the Perseus Molecular Cloud." *Monthly Notices of the Royal Astronomical Society* 523, no. 2 (2023): 2876–2886. https://doi.org/10.1093/mnras/stad1535.

International Theological Commission. "Communion and Stewardship: Human Persons Created in the Image of God." 2002, vatican.va.

———. "The Hope of Salvation for Infants Who Die Without Being Baptized." 2007, vatican.va.

Irenaeus. *Against Heresies*. Translated by John Keble. Oxford: James Parker, 1872.

———. *The Demonstration of the Apostolic Preaching*. Translated by John Behr. Crestwood, NY: St. Vladimir's Seminary Press, 1997.

John Paul II. “Cosmology and Fundamental Physics.” Address to the Pontifical Academy of Sciences, October 3, 1981, https://www.ewtn.com/catholicism/library/cosmology-and-fundamental-physics-8135.

———. “On Evolution.” Message to the Pontifical Academy of Sciences, October 22, 1996. *Origins* 26, no. 25 (December 5, 1996): 415.

———. *Fides et Ratio*. Encyclical letter, September 14, 1998, vatican.va.

———. “Letter of His Holiness John Paul II to Reverend George V. Coyne, SJ, Director of the Vatican Observatory.” June 1, 1988, vatican.va.

Kemp, Kenneth. “A Brief History of Catholic Evolutionism.” Society of Catholic Scientists, May 1, 2021, https://catholicscientists.org/articles/a-brief-history-of-catholic-evolutionism/.

———. “God, Evolution, and the Body of Adam.” *Scientia et Fides* 8, no. 2 (2020): 139–172.

———. “Problems of Conceiving Human Origins.” *Church Life Journal*, August 10, 2020, https://churchlifejournal.nd.edu/articles/problems-of-conceiving-human-origins/.

———. “Science, Theology, and Monogenesis.” *American Catholic Philosophical Quarterly* 85, no. 2 (2011).

Kim, N.H., et al. “Real-time Transposable Element Activity in Individual Live Cells.” *Proceedings of the National Academy of Sciences of the United States of America* 113, no. 26 (2016): 7278–7283. https://doi.org/10.1073/pnas.1601833113.

Leo XIII. *Providentissimus Deus*. Encyclical letter, November 18, 1893, vatican.va.

Lewis, C.S. “Is Theology Poetry?” In *The Weight of Glory*, 116–140. San Francisco: HarperOne, 1949.

Long, J.A., and M.S. Gordon. "The Greatest Step in Vertebrate History: A Paleobiological Review of the Fish-Tetrapod Transition." *Physiological and Biochemical Zoology* 77, no. 5 (2004): 700–719. https://doi.org/10.1086/425183.

Low, Y., et al. "Transposable Element Dynamics and Regulation during Zygotic Genome Activation in Mammalian Embryos and Embryonic Stem Cell Model Systems." *Stem Cells International* (October 15, 2021): 1624669. https://doi.org/10.1155/2021/1624669.

Mallick, S., et al. "An Early Modern Human from Romania with a Recent Neanderthal Ancestor." *Nature* 524 (2015): 216–219. https://doi.org/10.1038/nature14558.

Maritain, Jacques. *Untrammeled Approaches*. In *Opera Omnia*. Translated by B.E. Doering, 10: 118–129. Notre Dame, IN: University of Notre Dame Press, 1997. https://inters.org/maritain-dynamism-nature-evolution.

Maximus the Confessor. *Ambiguum*. Translated by Paul Blowers and Robert Louis Wilken. Crestwood, NY: St. Vladimir's Seminary Press, 2003.

McDougall, Ian, et al. "Stratigraphic Placement and Age of Modern Humans from Kibish, Ethiopia." *Nature* 433 (2005): 733–736. https://doi.org/10.1038/nature03258.

McGrath, Alister. *Darwinism and the Divine: Evolutionary Thought and Natural Theology*. Hoboken, NJ: Wiley-Blackwell, 2011.

McMullin, Ernan. "Cosmic Purpose and the Contingency of Human Evolution." 1996. Pari Center website, https://paricenter.com/library/culture-philosophy-and-science/cosmic-purpose-and-the-contingency-of-human-evolution/.

Melamed, D., et al. "De Novo Mutation Rates at the Single-Mutation Resolution in a Human *HBB* Gene Region Associated with Adaptation and Genetic Disease." *Genome Research* 32, no. 3 (2002): 488–498.

Meneganzin, A., and A. Currie. "Behavioural Modernity, Investigative Disintegration & Rubicon Expectation." *Synthese* 200, no. 47 (2022), https://doi.org/10.1007/s11229-022-03491-7.

Miller, Kenneth. *Only a Theory: Evolution and the Battle for America's Soul.* New York: Viking, 2008.

Monroe, J.G., et al. "Mutation Bias Reflects Natural Selection in *Arabidopsis thaliana*." *Nature* 602 (2022): 101–105. https://doi.org/10.1038/s41586-021-04269-6.

Myers, P.Z. "Synteny: Inferring Ancestral Genomes." *Nature Education* 1, no. 1 (2008): 47. https://www.nature.com/scitable/topicpage/synteny-inferring-ancestral-genomes-44022/.

Newman, John Henry. *An Essay on the Development of Christian Doctrine.* Park Ridge, IL: Word on Fire Classics, 2017.

———. Letter of December 9, 1863. In Dwight Culler, *The Imperial Intellect*, 267. New Haven, CT: Yale University Press, 1955.

———. "Letter to Edward Bouverie Pusey," June 5, 1870. In *Letters and Diaries of John Henry Newman*, vol. 25, *The Vatican Council, January 1870–December 1871.* Edited by Charles Stephen Dessain. Nashville, TN: Thomas Nelson, 1961.

———. "Letter to J. Walker of Scarborough on Darwin's Theory of Evolution," May 22, 1868. https://inters.org/Newman-Scarborough-Darwin-Evolution.

Newman, Stuart A. "Form and Function Remixed: Developmental Physiology in the Evolution of Vertebrate Body Plans." *Journal of Physiology* 592, no. 11 (2014): 2403–2412.

Nielsen, R., et al. "Tracing the Peopling of the World through Genomics." *Nature* 541 (2017): 302–310. https://doi.org/10.1038/nature21347.

Noble, Denis, et al. "Evolution Evolves: Physiology Returns to Centre Stage." *Journal of Physiology* 592, no. 11 (2014): 2237–2244.

O'Callaghan, John. "Evolution and Catholic Faith." In *Darwin in the Twenty-First Century: Nature, Humanity, and God.* Edited by P. Sloan, G. McKenny, and K. Eggleson. South Bend, IN: University of Notre Dame Press, 2015.

Origen of Alexandria. *On First Principles.* Translated by Frederick Crombie. In Ante-Nicene Fathers, vol. 4. Edited by Alexander Roberts, James Donaldson, and A. Cleveland Coxe. Buffalo, NY: Christian Literature, 1885.

Ortlund, Gavin. *Retrieving Augustine's Doctrine of Creation: Ancient Wisdom for Current Controversy.* Westmont, IL: InterVarsity, 2020.

Owen, Hugh. "Hugh Owen's Response to Cardinal Ruffini on the Days of Creation." https://www.pcpbooks.net/hugh_owen_and_cardinal_ruffini.html.

Pascal. *Pensées.* New York: E.P. Dutton, 1958.

Paul VI. Address to Theologians at the Symposium on Original Sin. July 11, 1966, vatican.va.

Pinsent, A. "Augustine, Original Sin, and the Naked Ape." In *Finding Ourselves after Darwin: Conversations on the Image of God, Original Sin, and the Problem of Evil.* Edited by Stanley Rosenberg. Ada, MI: Baker Academic, 2018.

Pius X. *Praestantia Scripturae.* Motu proprio, November 18, 1907. *The Sources of Catholic Dogma.* Edited by Henry Denzinger and Karl Rahner, trans. R.J. Deferrari (St. Louis, MO: Herder, 1954), 543.

Pius XII. *Divino Afflante Spiritu.* Encyclical letter, September 30, 1943, vatican.va.

Pius XII. "God the Only Commander and Legislator of the Universe." Address to the Pontifical Academy of Sciences, November 30, 1941, https://www.ewtn.com/catholicism/library/to-plenary-session-of-the-pontifical-academy-of-sciences-8957.

———. *Humani Generis*. Encyclical letter, August 12, 1950, vatican.va.

Pomeroy, E., et al. "New Neanderthal Remains Associated with the 'Flower Burial' at Shanidar Cave." *Antiquity* 94, no. 373 (2020): 11–26. https://doi.org/10.15184/aqy.2019.207.

Pontifical Biblical Commission. "Concerning the Historical Nature of the First Three Chapters of Genesis." June 30, 1909. In *A Catholic Commentary on Holy Scripture*. Edited by Bernard Orchard and Edmund F. Sutcliffe. Toronto: Thomas Nelson, 1953.

———. "Regarding the Sources of the Pentateuch and the Historical Value of Genesis 1–11." January 16, 1948.

Pontifical Council for Justice and Peace. *Compendium of the Social Doctrine of the Church*. June 29, 2004, vatican.va.

Poznik, G. David, et al. "Sequencing Y Chromosomes Resolves Discrepancy in Time to Common Ancestor of Males Versus Females." *Science* 341 (2012): 562–565. https://doi.org/10.1126/science.1237619.

Ramage, Matthew. *From the Dust of the Earth: Benedict XVI, the Bible, and the Theory of Evolution*. Washington DC: The Catholic University of America Press, 2022.

Ratzinger, Joseph. *"In the Beginning . . .": A Catholic Understanding of the Story of Creation and the Fall*. Grand Rapids, MI: Eerdmans, 1995.

———. "Schöpfungsglaube and Evolutionstheorie." In *Wer ist das eigentlich—Gott?* ed. H.J. Schulz. Munich: Kösel, 1969. Reprinted in *Creation and Evolution: A Conference with Pope Benedict XVI in Castel Gandolfo*. Edited by Stephan Horn and Siegfried Wiedenhofer. San Francisco: Ignatius, 2007.

Ratzinger, Joseph. *Truth and Tolerance: Christian Belief and World Religions*. Translated by Henry Taylor. San Francisco: Ignatius, 2004.

Rendu, William, et al. "Evidence Supporting an Intentional Neandertal Burial at La Chapelle-aux-Saints." *PNAS* 111, no. 1 (2014): 81–86.

Rescher, Nicholas. "The Problem of Future Knowledge." *Mind and Society* 11, no. 2 (2012): 149–163.

Rich, T.H., et al. "Independent Origins of Middle Ear Bones in Monotremes and Therians." *Science* 207 (2005): 910–914.

Rosenberg, Stanley P. "Can Nature Be 'Red in Tooth and Claw' in the Thought of Augustine?" In *Finding Ourselves after Darwin: Conversations on the Image of God, Original Sin, and the Problem of Evil*. Edited by Stanley P. Rosenberg. Grand Rapids, MI: Baker Academic, 2012.

Ruffini, Ernesto. *La Teoria dell'evoluzione secondo la scienza e la fede*. Norcia: Orbis Catholicus, 1948. Translated by Francis O'Hanlon in *The Theory of Evolution Judged by Reason and Faith*. New York: Wagner, 1959.

Russell, Bertrand, and Frederick Copleston. "Debate on the Existence of God." 1948. Reprinted in *The Existence of God*. Edited by John Hick, 167–190. New York: Macmillan, 1964.

Saint Mary's Press Catholic Research Group and The Center for Applied Research in the Apostolate. *Going, Going, Gone: The Dynamics of Disaffiliation in Young Catholics*. Winona, MN: St. Mary's, 2017.

Sander, P., et al. "Biology of the Sauropod Dinosaurs: The Evolution of Gigantism." *Biological Reviews* 86, no. 1 (2011): 117–155. https://doi.org/10.1111/j.1469-185X.2010.00137.

Scheeben, Matthias Joseph. *Handbuch der katholischen Dogmatik*. Book 3, *Schopfungslehre*. Reprinted in Mariano Artigas et al., *Negotiating Darwin: The Vatican Confronts Evolution 1877–1902*. Baltimore, MD: Johns Hopkins University Press, 2006.

Schönborn, Christoph. "Creation and Evolution: To the Debate as It Stands." *Zenit*, October 2, 2005, https://www.ewtn.com/catholicism/library/cardinal-schnborn-on-creation-and-evolution-10246.

Shank, Michael. "That the Medieval Christian Church Suppressed the Growth of Science." In *Galileo Goes to Jail and Other Myths about Science and Religion*. Edited by Ronald L. Numbers, 19–27. Cambridge, MA: Harvard University Press, 2009.

Simpson, G.G. *The Meaning of Evolution*. New Haven, CT: Yale University Press, 1967.

Slimak, Ludovic, et al. "Modern Human Incursion into Neanderthal Territories 54,000 Years Ago at Mandrin, France." *Science Advances* 8 (2022). https://doi.org/10.1126/sciadv.abj9496.

Standon, E.M., et al. "Developmental Plasticity and the Origin of Tetrapods." *Nature* 513 (2014): 54–58.

Su, D.F. "The Earliest Hominins: Sahelanthropus, Orrorin, and Ardipithecus." *Nature Education Knowledge* 4, no. 4 (2013). https://www.nature.com/scitable/knowledge/library/the-earliest-hominins-sahelanthropus-orrorin-and-ardipithecus-67648286/.

Swamidass, S. Joshua. *The Genealogical Adam and Eve: The Surprising Science of Universal Ancestry*. Downers Grove, IL: InterVarsity, 2019.

Tabaczek, Mariusz. "Thomistic Response to the Theory of Evolution: Aquinas on Natural Selection and the Perfection of the Universe." *Theology and Science* 13, no. 3 (2015), 325–344.

———. "What Do God and Creatures Really Do in Evolutionary Change? Divine Concurrence and Transformism from the Thomistic Perspective." *American Catholic Philosophical Quarterly* 93, no. 3 (2019): 458–459. https://doi.org/10.5840/acpq2019514179.

Teixeira, J.C., and A. Cooper. "Using Hominin Introgressions to Trace Modern Human Dispersals." *PNAS* 116, no. 31 (2019): 15327–15332.

Tenesa, A., et al. "Recent Human Effective Population Size Estimated from Linkage Disequilibrium." *Genome Research* 17, no. 4 (2007): 520–526. https://doi.org/10.1101/gr.6023607.

Tertullian. *De paenitentia*. In *Treatises on Penance: On Penitence and Purity*. Translated by William P. LeSaint. Mahwah, NJ: Paulist, 1959.

Thomas Aquinas. *De Potentia Dei*. Translated by the English Dominican Fathers. Westminster, MD: The Newman Press, 1952.

———. *De substantiis separatis*. Translated by Francis Lescoe. Carthagen, OH: The Messenger, 1963.

———. *In II Sent. Aquinas on Creation: Writings on the "Sentences" of Peter Lombard 2.1.1*. Translated by Steven E. Baldner and William E. Carroll. Toronto: Pontifical Institute of Mediaeval Studies, 1997.

———. *Summa contra Gentiles*. Translated by Vernon Bourke. Garden City, NY: Image Books, 1956.

———. *Summa theologiae*. Second and revised edition, 1920. Translated by the Fathers of the English Dominican Province, newadvent.org.

Trent, Council of. Decree on Original Sin. June 17, 1546. *The Sources of Catholic Dogma*. Edited by Henry Denzinger and Karl Rahner. Translated by R.J. Deferrari. St. Louis, MO: Herder, 1954.

Tretyachenko, V., et al. "Random Protein Sequences Can Form Defined Secondary Structures and Are Well-Tolerated *In Vivo*." *Scientific Reports* 7, no. 15449 (2017). https://doi.org/10.1038/s41598-017-15635-8.

Tucci, S., and J. Akey. "A Map of Human Wanderlust." *Nature* 538 (2016): 179–180. https://doi.org/10.1038/nature19472.

Tylen, Kristen, et al. "The Evolution of Early Symbolic Behavior in Homo sapiens." *PNAS* 117, no. 9 (2020): 4578–4584.

Vatican Council I. *Dei Filius*. Dogmatic constitution, April 24, 1870, inters.org.

Vatican Council II. *Dei Verbum*. In *The Word on Fire Vatican II Collection: The Constitutions*, ed. Matthew Levering, 43–149. Park Ridge, IL: Word on Fire, 2021.

———. *Gaudium et Spes*. In *The Word on Fire Vatican II Collection: The Constitutions*, ed. Matthew Levering, 211–337. Park Ridge, IL: Word on Fire, 2021.

Villanea, F.A., and J.G. Schraiber. "Multiple Episodes of Interbreeding between Neanderthal and Modern Humans." *Nature Ecology & Evolution* 3, no. 1 (2019): 39–44. https://doi.org/10.1038/s41559-018-0735-8.

Ward, C.V., and A.S. Hammond. "Australopithecus and Kin." *Nature Education Knowledge* 7, no. 3 (2016). https://www.nature.com/scitable/knowledge/library/australopithecus-and-kin-145077614/.

Ward, Keith. *The Big Questions in Science and Religion*. West Conshohocken, PA: Templeton, 2008.

Webb, George. *The Evolution Controversy in America*. Lexington: University Press of Kentucky, 2002.

Welker, Frido, et al. "Palaeoproteomic Evidence Identifies Archaic Hominins Associated with the Chatelperronian at the Grotte du Renne." *PNAS* 113, no. 30 (2016). https://doi.org/10.1073/pnas.1605834113.

Wilson, E.O. *The Diversity of Life*. Cambridge, MA: Harvard University Press, 1992.

Wilson, G., et al. "A Large Carnivorous Mammal from the Late Cretaceous and the North American Origin of Marsupials." *Nature Communications* 7, no. 13734 (2016). https://doi.org/10.1038/ncomms13734.

Wurz, S. "The Transition to Modern Behavior." *Nature Education Knowledge* 3, no. 10 (2012). https://www.nature.com/scitable/knowledge/library/the-transition-to-modern-behavior-86614339/.

Index